Recipe for Survival

What You Can Do to Live a Healthier and More Environmentally Friendly Life

Dana Ellis Hunnes
University of California, Los Angeles

CAMBRIDGE
UNIVERSITY PRESS

CAMBRIDGE
UNIVERSITY PRESS

University Printing House, Cambridge CB2 8BS, United Kingdom

One Liberty Plaza, 20th Floor, New York, NY 10006, USA

477 Williamstown Road, Port Melbourne, VIC 3207, Australia

314–321, 3rd Floor, Plot 3, Splendor Forum, Jasola District Centre,
New Delhi – 110025, India

103 Penang Road, #05–06/07, Visioncrest Commercial, Singapore 238467

Cambridge University Press is part of the University of Cambridge.

It furthers the University's mission by disseminating knowledge in the pursuit of
education, learning, and research at the highest international levels of excellence.

www.cambridge.org
Information on this title: www.cambridge.org/9781108832199
DOI: 10.1017/9781108935340

© Cambridge University Press 2022

First published 2022

Printed in the United Kingdom by TJ Books Limited, Padstow Cornwall

A catalogue record for this publication is available from the British Library.

Library of Congress Cataloging-in-Publication Data
NAMES: Dana Ellis Hunnes, author.
TITLE: Recipe for survival : what you can do to live a healthier and more environmentally
    friendly life / Dana Ellis Hunnes, University of California, Los Angeles.
DESCRIPTION: Cambridge, United Kingdom ; New York, NY : Cambridge University Press,
    [2021] | Includes bibliographical references and index.
IDENTIFIERS: LCCN 2021024688 (print) | LCCN 2021024689 (ebook) |
    ISBN 9781108832199 (hardback) | ISBN 9781108940658 (paperback) |
    ISBN 9781108935340 (epub)
SUBJECTS: LCSH: Environmental health. | Human ecology. | BISAC: SCIENCE /
    Environmental Science (see also Chemistry / Environmental) | SCIENCE / Environmental
    Science (see also Chemistry / Environmental)
CLASSIFICATION: LCC RA565 .H86 2021 (print) | LCC RA565 (ebook) | DDC 613—dc23
LC record available at https://lccn.loc.gov/2021024688
LC ebook record available at https://lccn.loc.gov/2021024689

ISBN 978-1-108-83219-9 Hardback

I dedicate this book to my loving and ever supportive parents; my husband, who encouraged me throughout this endeavor; and my thoughtful and empathetic son, who was my inspiration in writing this book.

The greatest threat to our planet is the belief that someone else will save it.

—Robert Swan

CONTENTS

FIGURES

TABLES

FOREWORD

"When you're talking about losing all of nature, it's not a spectator sport anymore, everybody has to become active somehow."
– *Louie Psihoyos*

As a global society, faced with myriad environmental and humanitarian challenges in the twenty-first century, perhaps we would be wise to observe and learn from the American bison. In the Great Plains, when the skies turn dark, and the wind picks up, they don't panic and bolt in the opposite direction of the impending storm; they don't turn their backs to the "problem." Instead, together they walk toward the ominous sky as one, exhibiting their instinctual awareness that the shortest time in the storm is achieved by acknowledging it and facing it head-on.

When it comes to tackling the existential threat of climate change, protecting our food and water sources, and the air we breathe, our best defense would be to act with grit, courage, and intention. *Recipe For Survival* spells out both the "why" and the "how" we can achieve these goals as individuals, and as a world community. Dana Ellis Hunnes has done the heavy lifting for us, dedicating years of her life to researching a path forward that places solutions in the palm of our hands, and hope on our horizon. She has crafted a masterpiece of a guide where we discover useful, everyday strategies for living a healthier life, one that incorporates doing so in harmony with our flora and fauna.

Imagine for a moment being asked to endure a month with no food or water, or even just a week. We are quickly reminded of how our own health and well-being is intimately entwined with having access to these natural resources, our most basic needs. Today, the jury is no longer out, it is indisputable that we must become better stewards of our environment – if not merely for ethical reasons, most certainly for the purpose of ensuring our ability to exist. The more in-depth our understanding of how reliant we are on a clean and thriving ecosystem, the more likely we are to feel a sense of urgency to protect it – to do so as if our lives depend upon it, because they do.

As a member of the team at Oceanic Preservation Society (OPS, a nonprofit organization that produces films and projection events focused on critical environmental issues), and as an author and spokesperson for environmental education, I often witness the changing of hearts and minds when an audience trusts the messenger, and thus the message. World leaders, corporations, and the majority of private citizens are finally demonstrating an acceptance that climate science can and must be trusted, and recognize the data as a resounding wake-up call. We are a reflection of our planet, and our planet is a reflection of us. For far too long we have not wanted to acknowledge what that looks like; from plastic pollution to greenhouse gases, the collateral damage is not pretty. The massive, daily carbon footprint from seven billion people is crushing Mother Earth and threatens to destroy what has taken over 4.6 billion years to create.

I envision Earth as a living, breathing goddess, fiercely resilient, yet signaling that she is ailing, injured by our reckless actions dating all the way back to the Industrial Revolution. She knows how to get our attention by using her uniquely supreme language. We hear her cry in the form of hurricanes, floods, extreme droughts, tornados, sea level rise, record-breaking temperatures year after year, ocean acidification, diminished food supply, melting glaciers and layers of permafrost, massive wildfires, depletion of natural drinking water, and yes, even pandemics.

Fortunately, thanks to scientists worldwide, we have identified the driving forces for these extreme conditions, and we know what it will take to reverse them. The United Nations' Sustainable Development Goals provide a comprehensive blueprint for how to arrive at a sustainable future for all. Their seventeen major objectives are attainable if the international community acts as a united front. With 197 member states now signed-on to the Paris Climate Agreement, there is reason to be hopeful; the notion that we are all in this together just may be what gets us through this quagmire. No one debates any longer whether or not the world is flat, the same should be true about climate change and the need to address it. Future generations have the right to a habitable planet – hardly a radical concept – and it is our fundamental responsibility to deliver.

With *Recipe For Survival*, we find practical and prudent recommendations for how to get involved. Dana's expertise as a plant-based dietitian and environmentalist enables her to provide us with twenty-one compelling "recipes" for how best to ensure that humanity, all wildlife, and Mother Earth survive and thrive. Instead of being the problem, we can now choose to be the solution. Educating and empowering people to become part of the growing force for change is at the heart of Dana's life's work. And let us keep in mind that while education raises awareness, awareness is not a resting place; instead, it is merely a launching pad for what is most needed: action.

There is strength in numbers, especially when the voice of the people is activated around a call for action. Our voices matter. Whether we look to an inspirational voice like climate activist Greta Thunberg, or tens of millions that turn out for Climate Marches, all efforts help to move the needle. In the pages that follow, some of the best ways to exercise our voice are revealed. A little sneak preview: embracing a plant-based diet (a form of voting for climate action with your fork); holding politicians accountable at the ballot box; being conscious consumers in the marketplace. Since supply is dictated by demand, as consumers we have the upper hand. If we don't buy it, they won't try to sell it. This applies to everything from the sale of endangered animal products, to oil, coal, and natural gas.

As a mother of four, I am all the more passionate about my work as an environmentalist; but it is all children I care about when I show up each day to advocate for clean air, clean water, justice for all species, and more. Outside of the workday, I try to keep contributing to a better tomorrow. I drive an electric car that I charge via solar panels designed to power our entire home. I boycott plastic at every chance, a petroleum product that is choking our marine life and trashing this celestial body. If we lead by example, we tend to give others permission to do the same. Dana's gift to every reader is an in-depth guide to the many different ways we can all participate as problem solvers, while encouraging others to join – and by the way, welcome to the team.

While there cannot be a one-size fits all scenario for how to take action, since we all have different abilities and situations, we all can and must see that there is a lot worth fighting for. We have far more leverage as citizens and consumers than we tend to recognize, and it is time to put that power into gear. Humanity may be embarking on space explorations to Mars, landing rovers, launching robotic orbiters, racing to claim first footprints on planets 100+ million miles away; but regardless, we won't discover another planet with all the bounty of Earth. Nowhere else is life possible for the human race. Home is right here, but we have taken it for granted.

As we ponder how to proceed, perhaps it helps to remember that since the beginning of time, love has often been the answer. French oceanographer Jacques Cousteau perhaps said it best, "People protect what they love." May we be wise enough to allow love to inspire our intentions. With each sunrise we are given a new chance to better love ourselves, others, and all living beings with which we share this great Earth – our common home. And as we leave this planet to those who will come after us, may we keep in mind the words of Native American, Chief Seattle, "We do not inherit the earth from our ancestors, we borrow it from our children." By living a more sustainable existence, we exhibit greater respect for all those who come after us. *Recipe For Survival* is the compass we need to get us there.

Jennifer R. Nolan
Board Chair / Dir. Strategic Engagement
OPS - Oceanic Preservation Society

PREFACE

> When we look down at the Earth from space, we see this amazing, indescribably beautiful planet. It looks like a living, breathing organism. But it also, at the same time, looks extremely fragile . . .
> —Space Shuttle/International Space Station astronaut Ron Garan

This quote reminds me of my son, right after his birth. I looked down at him, seeing an amazing and indescribably beautiful, living, breathing person, someone who was so tiny (weighing not even 6 pounds) and so extremely fragile. All I wanted to do was to protect him and shield him from anything that could possibly harm him – that's what every parent wants.

I also remember, when he was barely three years old, we were at the beach one day, and he asked me, "Mama, is the ocean big?" I responded with, "Yes, the ocean is big." He then asked, "And is there is a lot of water?" I replied, "There is so much water, you cannot possibly see it all in your lifetime." I clearly remember silently adding in my mind, "or the destruction we [humans] have caused."

I bring up these memories not to be dramatic, but to frame and remind myself of the importance of what is at stake here – why I wrote this book. I want to keep my son safe, and yet that is an increasingly challenging goal with the threats that environmental degradation and climate change pose.

I gave birth to my son less than a year after I completed my PhD in public health, focused on climate change and food security – issues that weigh heavily on me. So, while on maternity leave, to keep my mind stimulated I watched many documentaries and read many *more* books on the environment. Some cast light on the illegal black-market trade in endangered wildlife, others addressed the deforestation of the Amazon rain forest, and still others laid out the devastating consequences of palm oil plantations around the world.

There were times when these documentaries were so overwhelming that my heart ached from watching them, knowing that the planet being left to my son is hurting.

This is the same heartache I felt after leaving Ethiopia following my research on climate migrants – individuals who had left their villages to come to cities because they were unable to grow enough food at home – where I learned so much about the many environmental and food security challenges facing poor populations around the world.*

I'll admit that when writing my dissertation, I wrote academically (numbers, facts, figures), the text scrubbed clean of emotion, for that's what you do: report numbers and facts. Yet, after having my son, I realized that what I saw, learned, and wrote about was far more than just numbers, and it needed to be heard by more than just a handful of other academics.

What I learned is that climate change affects everyone. It affects some more than others, but it *does* affect everyone, including our children and grandchildren.† Climate change and its repercussions cannot be limited to just academic and high-level policy discussions – which all too often seem to go nowhere. Instead, these issues have serious consequences for the habitability and sustainability of our planet and for life itself (we are already seeing too many extinctions that may turn into a mass extinction event). We should *all be engaged*.

So, I took the next logical step: I began writing and speaking about all that I had learned. I educated others. My mantra became "Learn. Do. Educate." And that is exactly what I continue to do. This book is part of that.

As a registered dietitian (RD) at a major medical center in Los Angeles, California,‡ I work with hundreds of individuals every year. Many of these individuals are sick with debilitating chronic diseases;

---

* My dissertation is available at https://escholarship.org/uc/item/5bn1j3vv.
† At the time of this writing, a cabin my husband and in-laws built with their own labor, blood, sweat, and tears burned down due to an inferno that resulted from serious drought conditions and climate change.
‡ UCLA Medical Center.

some are even waiting in hospital for an organ transplant. My job is to guide them toward a healthier lifestyle through actions they can do every day to reverse or reduce their risk for heart disease, obesity, diabetes, or cancer, and actually achieve better health.

I also teach a graduate course in the Fielding School of Public Health at the University of California, Los Angeles (UCLA), where I teach students about the importance and power of nutrition for their overall health and well-being. I give students the knowledge to help them be their healthiest selves. Throughout the course, we discuss issues surrounding health, nutrition, and the environment. I am always surprised that although these students study public health, many of them were not previously aware that there are connections among the foods we eat, chronic disease, and the environment. If *they* don't know, how can I expect anyone else to?

Everything around us is interconnected. The foods we eat influence our health and also affect the environment. Our daily choices matter. That is what this book is about: Informing you on some of the issues we face (Part 1 of this book) and providing you with a number of ideas, solutions, or "recipes" of hope (Part 2 of this book) that can improve your own health and well-being, and at the same time improve the health and well-being of the place we all call home: Earth.

I spend my days working with individuals, and when I see them months or years later, I am often pleased at how much healthier they tell me they feel and how they had not previously realized how important nutrition is to their health and vitality. To be fair, I do see some of the sickest patients in the world. I give them knowledge they can use to be more empowered about their health, and in this book I give you information and tools you can use to keep *yourself* healthy and also help the environment.

I'll be the first to admit that 2020 was a bleak year, with the COVID-19 pandemic, news about climate change, and other environmental and sociopolitical problems leaving us to wonder if there is anything we can do. The answer is *yes*. We can always do something to improve our health and the environment around us, and to that end I provide in this book twenty-one "recipes" *you* can follow, starting today.

We can and we must find the will to take action, just like a doctor would with a patient. In fact, if Earth were a patient, I imagine we would do everything possible to save her. We would give her oxygen to breathe, remove excess carbon dioxide from her lungs and blood, and decontaminate or detoxify her from the poisons and toxic sludge flowing throughout her circulatory systems that have been building up over the years. We would cool down her fever and provide treatments to rehabilitate her after the years of harm she has experienced. We would do everything possible to help her live a healthy and long life and recover from what ails her.

Yet, I see too often that we do not treat Earth as though she is just such a patient who desperately needs life support. In fact, we are not treating Earth at all. Instead of replenishing her fluids, we dehydrate her land by pumping out too much water and melting the vast ice sheets that help keep her cool. We continue to harm her with more toxins – including plastics, fertilizers, pesticides, carbon dioxide, methane, nitrous oxide, and other greenhouse gases – every year. We tear down her cleansing mechanisms – trees, algae, phytoplankton, and other plant life. We push her closer to the point at which her heartbeat (the oceans), lungs (the rain forests and oceanic algae), and circulatory system (the rivers and streams) will simply *stop* working as they should. In their place we find beached whales, dead sea turtles, dead birds, burned forests, toxic algae blooms, and ocean dead zones.

In fact, if Earth were a patient, we would be *failing* her, and failing to do all we can to save her by giving her the lifesaving treatments and therapeutics she so desperately needs.

What if patient-Earth was your mother or your father? What if she was your sister or your brother? What if she was your daughter or your son? You would probably be upset or even horrified. You would probably wonder why the health care system was failing and just letting them die. Sadly, this is what we are doing to Earth. We are watching her and our fellow Earthlings die. But we *do* have the power to change this and end the destruction. We *do* have the power to change the trajectory of this sick patient, the Earth. That's what this book is about: How *we* can help make the planet and – at the same time – ourselves healthier.

This is what I do every day as a dietitian. I teach patients, many of them very sick, how to eat for health – how to eat to reverse and prevent chronic and debilitating diseases. I teach them how to eat for life. This is also what I do as an environmentalist and someone who studies climate change. I teach about habits and diets that are sustainable and that protect the land, save water, and minimize greenhouse gas emissions (despite a growing human population). Finally, this is what I do as a mother: I sound the alarm and work to protect the planet for my son and future generations.

We face a number of critical dangers – many of them we can see *now*, and many of them are yet to come. I wrote this book to highlight these dangers and provide a set of recipes, of opportunities we can all take *right now* to slow down and reverse the trajectory we are on.

We only have one home, our Earth, just as we only have one body, our own. Just like we want doctors, dietitians, and nurses to do everything they can to help us when we are sick, we must do everything possible to save our Earth, for she is sick.

I felt an image of Earth was crucial to have on the cover of this book – it is our only home; and when Earth was first seen from space and its image captured, it inspired the environmental movement and may be "the most influential environmental photograph ever taken."[§] At the bottom of this page, I share the image that has further inspired my environmental movement: my son (Figure o.1).

Figure o.1 My son staring out at the ocean as a young boy

[§] Galen Rowell, US photographer.

All we have ever known is on this planet, one tiny little "blue-and-white ball, orbiting a rather normal star, tucked away on the outer edge of a galaxy,"** amid billions of other galaxies in a vast universe. She is in crisis, and only *we* can save her.

** Jim Lovell, US astronaut on the Apollo missions.

ACKNOWLEDGMENTS

First and foremost, I will start by thanking my amazing parents, Susan and David Ellis. They were pivotal in helping shape my life and my ideals during my formative years. They taught me right from wrong, gave me the space to speak about my beliefs, and really listened. They taught me not to be afraid to pursue or do the right things, the just things, even if they are hard. I greatly appreciate that they read my early drafts, helped me with some of my initial editing, offered suggestions for the cover image, and sent me articles and relevant images for use in my book. I thank you, Dad, for your copyright and lawyerly help, and I thank you, Mom, for your unwavering belief in me. I thank you both for sending me very timely and relevant reading suggestions. You both have always been my biggest cheerleaders in life by supporting my goal of making a difference in this world. Thank you both so much for your unconditional love and support on this project.

Thank you to my husband Eric – truly, the best husband ever – for giving me the time and the space to write (over many weekends) and for rearranging our house a number of times to give me that space. Thank you, Eric, for supporting me in this endeavor by entertaining our son and keeping him occupied when I couldn't. I must say, your salad-making skills have greatly improved throughout my writing. Just as my parents read the first drafts, you read the very final draft, and though your comments were few, they were very meaningful and made the book even better. I am grateful.

Thank you to my son, Hugh, for being my inspiration to write this book. Your goodness, your warm heart, and the way you care about the world and animals made writing this book so very much easier, because I know that you will appreciate and understand what I am writing it for. Thank you for taking a general interest in the topics I was writing about. I love that you engage me in conversations about each topic and ask a ceaseless number of questions, making me think more deeply about the importance of each topic. Even at six years old, your understanding and compassion for the variety of topics I wrote about astounded me. I know it has been a very challenging year, and I am beyond impressed with the little man you are growing to be. Thank you for supporting me in this journey, for entertaining yourself at times, and for being so caring. I wrote this book for you.

Thank you to my dog, Jack – sixteen years old at the time of this writing, and the best dog in the world – for teaching me what it is to be there unconditionally for the people you love and cherish, and for reinforcing in me that dogs – and frankly all animals – think, feel, and love, and that giving up is just not an option.

Finally, thank you to my acquisitions and copy editor, Matt Lloyd, for being willing to take a chance on this book. Back when I first came up with the idea of writing this book six years ago, the issues I discuss here were more radical and progressive, and there were few books on these topics. So, a big thank you to Matt for your confidence in my research and writing, and for taking an interest in these incredibly important ideas. Thank you also to Sarah Lambert, Sapphire Duveau, Divya Arjunan, and John Marr for reading through my manuscript, guiding me through the editing process, and ensuring that all would go smoothly with the transition from computer to book.

INTRODUCTION TO PART 1

Those who do not learn from the past are doomed to repeat it.
—George Santayana

I admit, writing Part 1 of *Recipe for Survival* was very difficult for me. It contains many issues that are disturbing, tragic, and perhaps even scary – issues that have kept me up at night. It may be upsetting and at times difficult to read, but it is incredibly important that you *do* read it. Part 1 (Chapters 1–7) describes some of the problems we face – primarily related to climate change and human behaviors – and it sets up the rationale and the need for Part 2 (Chapters 8 and 9), which describes hope and the actions *you* can take.

Our consumption habits and our competition for resources (often against other species) have had all kinds of negative impacts on our own health and on the world at large. We are now at a pivotal point in time – what we do right *now* matters more than *ever*.

This has been especially true in 2020, when business as usual was completely uprooted with the emergence of a deadly global pandemic. Many countries around the world forced their citizens into lockdown, confining them to their homes or apartments, where life as we all knew it stopped.

In over a year (as of May 2021), 161 million individuals have tested positive for COVID-19 and over 3.3 million individuals have died from it.

Too many people have died from it.

Even now, as I finish writing this book, we are still in the middle of the COVID-19 pandemic, and many people are struggling to survive this harrowing time. The number of hungry and food-insecure individuals around the world grew (again) during the pandemic.[1]

While some countries lack the health care infrastructure that is needed to save individuals, other countries have far too many individuals living with preexisting conditions, malnutrition, or who are negatively affected by climate change – compounding the effects of this disease.

It is possible that the COVID-19 pandemic could have been prevented. The virus is believed to have crossed over from a wild animal to a human in a wildlife trade market in Wuhan, China. This is not the first time something like this has happened. Nor will it be the last.

The flu of 1918–1919 was believed to have originated in birds. It crossed over into humans and infected 500 million individuals around the world: it killed 50 million.[2] This flu was so infectious and deadly to humans in part because we had no immunity to it.

We have no immunity to COVID-19 either.

Even more recent than the flu of 1918–1919 was the SARS-CoV-1 epidemic of 2003, which spread to more than two-dozen countries and killed 774 people. This virus (which was also a coronavirus) originated from close interactions between humans and wildlife.[3] Similarly, MERS – the Middle East respiratory syndrome of 2012 – is another coronavirus illness that was new to humans and likely originated in bats. MERS killed 858 people.[4,5]

The reason for bringing these examples up is that all of these viruses originated in wild animals, all of these viruses crossed over to humans as a result of close interactions, and all of these viruses were (and are) lethal. Moreover, if past is prologue, these examples tell us that there will almost certainly be viruses in the future that will cross over from animal reservoirs to humans – and cause disease – in a world with worsening climate change, habitat loss, human population growth, and increased human–wildlife interaction.[6]

This is one of the topics I explore more fully in this book – how humans exploit animals and the environment and what this may mean for our own health and for the health of the planet with its interconnected ecosystems. While this discussion is not for the faint of heart, it is necessary to have for a better understanding of the importance of each species and why we must not take any of them for granted.

There are other lessons we can learn from our experiences with the COVID-19 pandemic that apply to the ways we address environmental problems and climate change.

When it came to the COVID-19 pandemic, we were not prepared. We were left scrambling in a panic for personal protective equipment (PPE), testing, treatments, vaccines, and a cure. With climate change, we *do* have advance warning, and we have *had* advance warning for some time now.

We know global temperatures are rising. We know pollution is more prevalent. We know carbon emissions are increasing. And we know glaciers are retreating. We see the polar regions losing their ice. We see species fleeing their homes or dying where they cannot adapt fast enough to environmental changes.

We have the signs. We see the symptoms. If we are not careful, we will hit *the* tipping point, when it will be too late to protect or save what we have here on Earth. These are topics of discussion throughout Part 1 of this book.

We are also at a point – *right now* – when we still have time to *prevent* these things from happening (or at least lessen their potential effects) with sufficient foresight and will. That is what Part 2 of this book presents – solutions and life support.

Our planet needs us to step up and help it. I believe we can do it. I believe we can be empowered to actively heal the planet. It *is* possible. We have seen signs of Earth's healing and glimpses of recovery during the COVID-19 pandemic. During the lockdowns, Earth was given a much-needed respite from the daily onslaught of carbon emissions that result from cities.

In the short amount of time in which human activity was put on pause, Earth's respiratory and water systems began to show some signs of healing.

The world took a breath.

There were reports of cleaner canals throughout Venice; animals were seen swimming where none had been seen for decades. There were reports of cleaner air in China and Los Angeles where smog is the norm. There were even reports that where countries were in full lockdown there was an average decrease in energy demand of 25 percent. Countries that were in partial lockdown reported a decrease in energy demand of 18 percent. In fact, lockdowns around the world have lowered total global energy demand by up to 6 percent when looked at over the course of 2020. This is largest decrease seen in seventy years and the largest ever decrease in absolute terms.[7]

But, sadly, as with almost everything in life, there *is* a flip side.

The significant increase in the use (and discarding) of PPE has taken a major toll on the environment. While PPE is absolutely needed to protect healthcare workers and essential workers (and everyone, really) from spreading the coronavirus, far too much PPE – much of it made with plastic – has entered the oceans.[8] Gloves, masks, hand-sanitizer bottles, and more are thrown away every day by the million.[9] Most of these items are used once and then tossed in the rubbish pile.

There are also reports that land grabbing, deforestation, illegal mining, and wildlife poaching by opportunistic actors and criminal groups actually increased during the pandemic, adding stress to already fragile environments.[10,11]

What are we to make of these various changes, some good and some bad?

One lesson we can take away is that by using fewer resources and living more sustainably – especially by those of us who live in developed countries and use more than our fair share – we can help Earth slow down its warming trend, we can help fish stocks recover, and we can help rain forests to expand and draw down carbon dioxide from the atmosphere and put more oxygen into the atmosphere. We can allow wild species to thrive once again in their natural environments rather than being confined to small geographic areas or overcrowded cages where they may host and spread new deadly diseases.

During the pandemic, many of us saw and felt for the first time what it feels like to be cooped up in our homes day in and day out; to see the same surroundings twenty-four hours a day, seven days a week without much stimulation, mental or physical; to feel what animals confined in a factory farm, circus, zoo, or aquarium might feel. It has been emotionally taxing and mentally debilitating for many of us. And perhaps it has given us more insight into the way we treat others – whether animal or human. But we can prevent this from happening again in the future, whether from the next disease outbreak or from too many climate extremes. We must each do our part.

There are many examples where individuals performed acts of kindness or gave to others to help save lives during the COVID-19 pandemic. We must act in kind, each of us performing acts of sustainability, to help heal Earth. Although Part 1 of this book describes the challenges we face, Part 2 provides *hope* and recipes for how we each can help Earth and ourselves. So, let's dig in.

1 AGRICULTURE IS A MAJOR DRIVER OF CLIMATE CHANGE (AND DISEASE)

Our most basic common link is that we all inhabit this small planet. We all breathe the same air. We all cherish our children's future. And we are all mortal.
    —John F. Kennedy

Agriculture's Contribution to Climate Change and Chronic Disease

Climate change is experienced as long-term changes in temperature, precipitation, humidity, and wind speed. Climate change is predicted to negatively affect the productivity of agricultural, forestry, and fisheries systems in the years ahead because of an increase in the frequency and magnitude of weather extremes such as storms, floods, and drought.[††]

Climate variability is experienced as shorter-term seasonal shifts, which often have even more immediate and potentially deleterious effects on local environments.[12,13]

Individuals, farmers, pastoralists, and aquaculturists depend on the stability of the environment for their livelihoods and food security (as do large industrial farms, which often feed large numbers of people). The accelerating rate of climate change and the increased frequency of

---

[††] We are already seeing this.

negative environmental events and their ensuing damage are lowering the levels of growth, development, and yields within certain food systems – especially those near the equator, which historically have been very productive – and may push food-producing systems beyond what is recoverable.[12,14]

One of the regions that is most vulnerable to climate change and weather extremes is sub-Saharan Africa, which has the highest prevalence of food insecurity in the world. Latin America, the Caribbean, and south Asia are close behind.[12] Throughout these regions, poverty is endemic due to limited access to capital, infrastructure, and technology. Complex governance and institutional dimensions, conflicts, disasters, and disproportionate population growth also interact to contribute to these high poverty levels. Climate stressors and climate extremes further increase the likelihood of adverse health and food security outcomes.[15,16]

In a typical year, there are more than 800 million people around the world who suffer from malnutrition, hunger, or food insecurity because they are unable to grow or obtain enough food to eat. There are also 150 million children each year who present as underweight or stunted – which is a long-term indicator of poor nutritional intake.[17] Many people who are hungry or are at risk of starvation resort to extreme measures.[18] These can include selling important resources, poaching, or even migrating to feed themselves and their families.[19,20,21] Clearly, climate change and climate extremes have the biggest impacts on those who are the most vulnerable.

At the same time, there are 1.9 billion people around the world who are overweight or obese, who consume far *more* calories than they need for good health, and who live with a range of debilitating chronic diseases.[22]

Each year, the average Australian, American, or European consumes more than 75 pounds (34 kg) of meat (beef and pork), more than 57 pounds (26 kg) of poultry, at least 250 eggs, and at least 275 pounds (125 kg) of milk (primarily as cheese, frozen-dairy desserts, cream products, and fluid milk).[23,24] Most if not all of these foods have been linked to an increased risk for preventable chronic diseases such as diabetes, heart disease, stroke, kidney disease, and cancer.[25,26,27]

In summary, people who eat more animal products tend to have more chronic diseases. People who eat more animal products also contribute far more to climate change than those who do not have enough to eat. Diets that are high in animal proteins contribute significantly to environmental degradation and climate change.[28,29]

Did *you* know that?

I ask because even I, a dietitian who studies climate change, food security, and agriculture, was shocked to learn that the foods we eat produce between 25 percent and possibly as much as 51 percent of all greenhouse gas emissions globally.[29,30,31] Agricultural emissions are composed primarily of methane (chemical symbol CH_4), nitrous oxide (N_2O), and carbon dioxide (CO_2). It is worth noting that CH_4 is between 28 and 36 times stronger than CO_2 in terms of its warming potential; and even worse, N_2O is 265–298 times stronger than CO_2.[32,33]

Each year, livestock produce at least 7.1 gigatons of greenhouse gas emissions (and this is growing). That is more than the entire global transportation sector.[30,31] Up to 80 percent of these emissions are produced by the enteric fermentation of ruminants – namely beef and dairy cattle – and by the off-gassing of manure or other by-products of animal agriculture.[‡‡] Pigs, chickens, buffalo, and other small ruminants contribute to these emissions as well, but to a lesser degree.[34,35]

Every day we have the opportunity to choose the foods that we eat, and whether or not they will contribute to climate change. Sadly, this is rarely discussed at high-level meetings or in the news despite the fact that our food choices affect livelihoods, food security, sustainability, and health.

There is more than enough food grown around the world to feed every person a sufficient number of calories for good health. So much food is produced that nearly one-third of all food calories grown ends up getting discarded as food waste.[36] The United States produces nearly double the number of calories (3,800 per person per day) than are needed for good health (much of this food is sugary, processed, or animal-based). In contrast, in developing and low-income countries, calories are lacking. Clearly, the foods we produce are not well distributed, as evidenced by the large number of people who are hungry, the

---

[‡‡] Enteric fermentation is a digestive process in cows and other ruminant animals (cattle, sheep, goats, buffalo) in which carbohydrates (such as grasses or other grains) are broken down by microorganisms in the stomach (rumen) or digestive tract into simple molecules (such as glucose) for absorption into the bloodstream of the animal. Unfortunately, this process produces methane as a by-product.

even larger number of people who are overweight or obese, and the amount of food that is thrown away.

Food waste – which occurs at the farm and in the home – is another large contributor to greenhouse gas emissions. As much as 40 percent of food calories get discarded (depending on the country), and food waste produces 7–8 percent of all greenhouse gas emissions, primarily from methane – the by-product of all that waste. If "food waste" were its own country, it would be the third-largest emitter of greenhouse gases in the world, after the United States and China. The annual global emissions produced by food waste exceed the total annual emissions produced by all of the cars in the United States.[37]

So, not only is food waste a huge contributor of greenhouse gas emissions but also it represents a considerable loss of nutrients and calories from the worldwide food system.

Seeing this waste – this inability to move food to where it is needed, whether in developed or developing countries – is heartbreaking for several reasons including: the pointless waste of animal lives, the nutrients that are wasted from crops, the waste of energy resources that went into producing the food, and the hundreds of thousands – if not millions – of hungry or undernourished people who could have been fed this food.

The fact that so much food gets wasted highlights the many individuals and groups around the world who are vulnerable to malnutrition and food insecurity.

Further compounding these issues, roughly 42 percent of all cereal grains grown around the world are diverted away from human consumption to feed animals.[38] Yet in many parts of the developing world there are shortages of calories and micronutrients (vitamins and minerals) available for human consumption, resulting in mass malnutrition, disease outbreaks, an increase in risky behaviors such as poaching endangered animals, migration (most often as a last resort), or some combination of these.[39]

Conversely, among many higher-income nations, including the United States, Canada, and several European countries, nutrition-related chronic diseases represent *the most numerous causes of death*. Several studies point to the overabundance and overconsumption of animal products such as meat, poultry, dairy, and processed foods as the most likely proximate causes if these diseases.[40,41,42]

At both ends of the spectrum, individuals live with debilitating conditions – whether infectious or nutrition related – and too often die younger than they should.

Agriculture's Contribution to New Infectious Diseases

Another factor that contributes to both the initiation and spread of new infectious diseases and environmental degradation is the increasing prevalence of monoculture farming methods.

"Monoculture farming" is when only a single crop or organism is grown or cultivated on agricultural or forested land. A monoculture system differs from traditional farming, in which three, four, or even more different crop or animal species are grown or raised simultaneously. Monoculture farming, which has become increasingly common, is akin to putting "all of your eggs into one basket" and hoping for the best. This is quite risky.

A major problem associated with monoculture systems is the increased susceptibility of the crops or animals to disease. This is in part due to the lack of genetic diversity within the system itself, eliminating much of the resilience that species might have had in a more diverse setting with other species surrounding them to spread (and lower) the risk. It also means there is no line of defense in case of a disease outbreak.

Throughout the world, particularly in the tropics – which historically have been rich in biological diversity – rain forests are being razed and replaced with monocultures that are completely devoid of biological diversity. Monocultures are often paid for by giant agribusinesses and are the leading causes of biological diversity loss as they remove, displace, or simply kill native species.[13,43,44,45]

Like crops, many "industrial" animal farms today are also monocultures where one species with one specific genetic preference is raised. These animals are often kept in what is known as concentrated animal feeding operations (CAFOs).

CAFOs frequently use enclosed warehouse-type structures to house or confine their animals in extremely high density – thousands of large animals or hundreds of thousands of smaller animals.[§§] CAFOs

---

[§§] Cows are considered large animals, while chickens, ducks, sheep, or goats are considered small animals.

offer minimal, if any, mental stimulation, comfort, or opportunity for the animals to engage in natural behaviors.[6] They do, however, provide conditions that increase the likelihood that animals will acquire infectious pathogens and share those pathogens with one another, and so increase the chances of the pathogens evolving into new forms, some of which may be capable of infecting humans as well.[6]

CAFOs also often lead to poor farm management, environmental pollution and degradation, and public health problems, including diseases – both chronic and infectious. CAFOs may also have negative socioeconomic impacts on the farmers who work there and on the surrounding communities.[46,47,***] It should be noted that the primary objective of a CAFO is to maximize profit, not to worry about welfare.[47]

Today, CAFOs supply more than 90 percent of the world's meat and nearly 99 percent of the meat consumed in the United States.[48] Around the world, it is estimated that over 90 percent of the animals that are farmed live in factory farms. This includes both vertebrate land animals and farmed fish.[49]

To expedite the growth of their animals and mitigate the spread of disease or pestilence – which are common in crowded and stressful living conditions – CAFOs frequently give their animals prophylactic doses of medically important antibiotics and other drugs or pesticides. These treatments are "intended not to cure the animals, but to foster their weight gain and maintain their health just sufficiently for profitable sale and slaughter."[6,47]

In the United States, roughly 80 percent of *all* antibiotic use occurs on farms. By 2030, antibiotic use on farms will increase by 69 percent globally due to the expected growth in meat consumption in middle- and low-income countries.[50,51] This extraordinary use of antibiotics contributes significantly to the evolution of antibiotic-resistant bacteria in both animals and humans.[47,50]

Both our agricultural practices and our close contact with wild and domesticated species are projected to lead to new or emerging infectious diseases, and it is suggested that three out of every four new or emerging infectious diseases will come from animals.[44,52] CAFOs exacerbate the transmission of new diseases, for when diseases show up,

*** To see examples of CAFOs, head to any search engine and look up "CAFO images." What you will see may shock you.

they quickly spread through a herd or flock, causing significant harm (or death) to the animals. If the disease is *also* transmissible to humans, there may be few – if any – ways to treat it due to the decrease in the efficacy of currently available antibiotics, or simply due to the fact that it may be a disease that has never been seen before.[48,53,54]

This has happened a few times in the recent past, such as with the H5N1 bird flu in the early 2000s, the Middle East respiratory syndrome (MERS) in 2012, and even more recently the severe acute respiratory syndrome (SARS) coronavirus-2 (COVID-19) of 2019–2021.

The H5N1 bird flu is a highly pathogenic influenza virus that spilled over from wild birds to domesticated poultry for the first time in 1997.[6] It is highly contagious and deadly to wild birds and domesticated poultry. It is also extremely deadly to humans who become infected with it (up to 60 percent mortality rate) after direct contact with sick or dead poultry that were infected with the virus.[55] A vaccine has since been developed to help prevent the spread of H5N1 among birds; however, this vaccine is not given to humans, because as of yet H5N1 does not spread from human to human.[6]

The MERS virus of 2012 was a coronavirus that spread in the Middle East. MERS was thought to have originated in bats that transmitted the virus to camels, which then spilled over into humans who were in close contact with those camels.[56] MERS was also a new virus that spread from animals to humans, and it had a roughly 35 percent mortality rate. To date, there is no vaccine for MERS, but it did not rapidly spread around the world in the way certain flus or other coronaviruses have.

COVID-19 is a coronavirus that is believed to have spread from a wild animal reservoir, perhaps a bat, to humans by way of an intermediary animal that was housed in a market where there was close contact between animals and humans.[57] While this virus is not nearly as deadly as H5N1, SARS, or MERS, COVID-19 has seen widespread dissemination around the world and has killed more than 3.3 million individuals (as of this writing).

In addition to viruses that spill over from animals to humans, there are a number of bacterial infections that are spread by animals or in animal-based foods that cause illness or death in humans every year. The most common bacteria that cause foodborne illness in humans are *Staphylococcus aureus*, *Clostridium perfringens*, *Salmonella* species, *Escherichia coli*, and *Listeria* species, to name a few. The foods that commonly transfer these bacteria are of animal origin, in particular eggs and egg products, pig meat, broiler meat, cheese, fish and fish products, milk and dairy products, bovine meat, and crustaceans.[58]

Unfortunately, as more antibiotics are used in animals for nontherapeutic purposes, more bacteria have developed antibiotic resistance.[59]

Another factor that has contributed to the increased risk for new infectious diseases spreading from animals to humans is the rapid rate of growth in the human population over the last several decades. Unfortunately, this will likely worsen, as the human population is expected to continue to increase through at least 2050.[6]

Today, humans live in high densities in many cities around the world. We have penetrated ecosystems and encroached on forests, and we continue to do so into new wild areas around the planet. We disrupt the physical habitats and the ecological well-being of these places.[6] We kill and butcher and eat many of the wild animals that are out there. All of these behaviors increase the frequency of human–wild animal interactions and make species spillover of infectious diseases all the more likely.

Today's global population is roughly 7.8 billion people.[60] By 2100, the global population could be as high as 11.2 billion people. Population growth *is* slowing down in westernized, developed countries such as the United States, Canada, Japan, and several countries of Europe. It is expected that the majority of population growth will occur among several developing countries, specifically those in Africa, where populations are expected to double to more than 2.6 billion people (from today's 1.3 billion).[61,62,63] The population of India is set to reach 1.70 billion people by 2050, an increase of 25 percent from today's 1.38 billion, while the population of Bangladesh could increase by as much as 20 percent by 2050 (to close to 200 million people).[64,65]

A bigger population means more mouths to feed. More mouths to feed means more land is needed for growing more food. Clearly, more efficient and productive farming methods and changes in food choices will be needed.

In addition to population growth, where people live is also changing. Currently, around 55 percent of the world's population lives in cities. By 2050, a little more than 66 percent of the world's population is projected to be living in cities.[66] Along with urbanization often comes a diet containing more processed foods and more animal-based foods.[67] This phenomenon was seen in China. Starting in the 1970s, over a forty-year period, as incomes grew and the population urbanized, per capita meat intake in China quadrupled.[68]

It is projected that among African nations, meat intake will increase by 50 percent over the next fifteen to thirty years as a result of increased urbanization and increased incomes.[66] For these populations, meat and other animal products are considered more desirable and reflect a higher standard of living or wealth. They also frequently represent a "better nutritional intake."

This is something that struck me when I was in Ethiopia and engaged in my dissertation research. I met with individuals who had migrated from their villages into the city to earn a cash income, often for the following reasons: to be able to purchase more and better foods for themselves and their families; to help their families make money to buy oxen, irrigation, or other farming inputs; to purchase supplies for schooling and education, including pencils, books, and paper; or for all three of these reasons.[69] I was recently reminded of this while watching the film *Black Gold*, a documentary about coffee-farming regions and coffee farmers in Ethiopia struggling to survive and feed their families and to provide for schools and other education facilities.[70]

But if every person in the world started eating like an American, consuming as much meat and other animal products as Americans do, we would likely require a second Earth to feed us all![71,72] Unfortunately, we do not have a planet B to use for this purpose.

With this in mind, I find myself asking: How and where will this increase in meat come from, and what will happen to the remaining wildlife and their habitats in these countries, or in other countries where this meat is imported from?

Each year, 8 million hectares (nearly 20 million acres or 80,000 km²) of rain forest are destroyed for agriculture – for both subsistence and industrial purposes – and for cattle ranching.[73] This clear-cutting of forests removes oxygen from the atmosphere and releases millions of tons of CO_2, CH_4, and nitrogen, escalating climate change and biodiversity loss, increasing the potential for negative impacts on the health and sustainability of our planet.

In the world's largest rain forest – the Amazon – 88 percent of the loss that occurs is due to commodity-driven deforestation or clear-cutting for agriculture. Where forest once was, soy or other grains destined for animal consumption now grow.[44,73] The loss of rain forests has significant effects on the environment, both locally (by affecting local weather patterns) and more globally (in terms of regulating oxygen and CO_2 levels in the atmosphere). The loss of rain forests intensifies soil erosion and desiccation (drying out), reduces or completely removes forest foods and services, affects food supply, stability, and availability, and often results in the loss of unique species.

The loss of rain forests also pushes large numbers of food-insecure people onto plots of land that are too small, isolated, marginal, or unproductive. This means they have difficulty meeting their livelihood and food security needs. This is further evidence that the ways in which we currently grow, raise, or distribute our foods is not working.[74,75]

Currently, 13 percent of the Earth's land surface (1.7 billion hectares, 17 million km²) is too cold for rain-fed agriculture, and 27 percent (3.6 billion hectares, or 36 million km²) is too dry. Only 17 percent of the world's agricultural land is irrigated, but it produces 40 percent of all cereal crops. Desertification is expected to increase during this century, meaning there will be *less* arable land to grow food on.[76,†††] Yet, as the human population grows, cereal production will need to increase by 38 percent over current levels by 2050. Meat production will also likely increase by 62 percent (up to 573 million tons per year, or 520 million tonnes per year) due to the growing desire for animal products in the developing world.[77,78]

Unfortunately, it is wildly inefficient to give calories to animals to produce animal protein foods for humans. For every 100 calories of plant matter fed to animals, only about 3–17 percent of those calories turn into meat, dairy, fish, or egg proteins for human consumption.[79] Every year there are hundreds of millions of people throughout the world who are hungry or starving. Yet billions of animals are fed the calories and nutrients that *could be fed* directly to people in order to nurture a growing human population.[80]

Clearly, raising more animals for food is not the solution. As stated earlier, animal products increase the risk for many chronic diseases, and they also decimate the environment, pollute water, and produce too many emissions.[27,31,34,81]

---

††† Desertification is the process by which fertile land (often previously rain forest) becomes desert or nonarable land, typically as a result of drought, deforestation, or inappropriate agriculture. Up to 20 percent of the world's land is threatened with desertification.

A study from Cornell University in New York found that an animal-heavy diet, such as a standard American diet, requires significantly more land, water, and energy resources than a primarily plant-based diet.[82] Another study published in the *Proceedings of the National Academy of Sciences* indicated that, by 2050, a plant-based diet could lower food-related greenhouse gas emissions by between 29 and 70 percent as compared to an animal-heavy diet.[34]

These statistics underscore the importance of significantly lowering animal-foods consumption in higher-income nations *and* of keeping animal-foods consumption low in lower-income nations. Yet as people in developing countries urbanize and gain more disposable income, they will tend toward dietary patterns that mimic current "Western" dietary patterns that are heavy in animal products.

It is necessary that we balance the needs of those who are most vulnerable to hunger and malnutrition – and who can now afford more animal products – with the wants of those who use more than their share, as well as with what the planet can provide in a sustainable way. Achieving this balance will be a huge test of policy, science, and morality.

The more we exacerbate climate change and the loss of natural ecosystems and biological diversity, the more difficult it will be – and the less time we will have – to achieve these goals and this balance. At present, we seem to be collectively hurtling toward a cliff edge and someone has cut the brakes.

Many people already understand that fossil fuels are a major source of greenhouse gases, adding carbon to the atmosphere and oceans. However, far fewer people are aware that what we eat – animal products specifically – contribute enormously (even more than all of the cars, planes, and ships in the world) to climate change, sea-level rise, ocean acidification and dead zones, and global extinction rates.[83]

Low-lying island nations such as the Solomon Islands and the Maldives are disappearing under the ocean. These nation-states are being displaced and their people are clamoring for resources, including new places to live.

Coral reefs, the largest living masses on Earth, are dying and losing the sea creatures that call them home. Oceans are losing their diversity. Land animals are losing their habitats. Earth's ecosystems are being disrupted in what is being called the "sixth mass extinction." The first five mass extinction events were caused by natural occurrences; this time, the cause is us.[‡‡‡]

Yet we can delay and perhaps even reverse these calamities. In fact, one of the most impactful things we can do to reverse course and restore our global ecosystems is to significantly reduce our intake of animal proteins and significantly increase our intake of a whole-foods, plant-based diet.

Diet is "the most important contribution every individual can make to reversing global warming."[84] It is not *just* about how much we drive or how *much* electricity we use. Over the next decades, renewable energy use will likely increase and greenhouse gas emissions from transportation and energy will likely decrease as countries (hopefully!) band together to reduce their use. In contrast, emissions from agriculture, food waste, and animal husbandry will almost certainly increase due to population growth and demand, unless we do something about these as well. These foreseeable changes are why diets and *what we eat* need to be discussed – immediately – in climate change talks, in books, on the news, and at conservation and sustainability meetings.

The other major reason for why we need to have these discussions is the increased risk of new infectious disease outbreaks related to how we raise animals for food and how these new infectious diseases may act as effect multipliers of a person's vulnerability to climate change and food insecurity.

Infectious Diseases as Effect Multipliers of Climate Change and Food Insecurity

Vulnerability refers to the propensity that human livelihoods and assets will suffer adverse consequences when impacted by hazards such as climate extremes, changes to the food supply, or economic shifts that affect the ability to access foods.[18] While climate change and environmental stressors can obviously affect food production systems

---

[‡‡‡] The first five mass extinction events include the Ordovician (444 million years ago), when 86 percent of species were lost due to a severe ice age; the late Devonian (375 million years ago), when 75 percent of species were lost due to oceanic algal blooms; the End Permian (251 million years ago), when 96 percent of species were lost due to a cataclysmic volcanic eruption near Siberia; the Triassic (200 million years ago), when 80 percent of species were lost due to unknown reasons; and the Cretaceous (66 million years ago), when 76 percent of species were lost due to a gigantic asteroid impact.

and supply chains, nonenvironmentally related events such as pandemics can also affect food production systems and supply chains, particularly among those who are most vulnerable. This is something we have seen during the COVID-19 pandemic, which has increased the risk of malnutrition and even starvation for an additional 265 million people around the world. This is on top of the typical 800 million people globally who are malnourished every year.[85]

Social and political factors associated with the pandemic have interacted with environmental changes, food production systems, and supply chains to increase the vulnerability of certain populations to waning food supplies and malnutrition.

Any shock to a system, whether from climate change or a global pandemic, may result in both immediate and longer-term effects on food security. These changes may make some or many food products inaccessible or unaffordable to vulnerable populations. What this means for long-term food security in a world with a growing and more vulnerable population remains to be seen.[86] What is known, however, is that hungry and resource-stressed populations tend to mass migrate or to fight wars.

Food shortages – which occur more often in developing countries – have been exacerbated by the pandemic. Supply-chain problems have led to food shortages and/or have increased the price of foods. COVID-19 is an "effect multiplier." It has multiplied the stressors people already face from climate change, soil erosion, salinization, overgrazing, overextraction of groundwater, and the general decreased productivity of the land. COVID-19 has made things significantly worse for many people around the world.[87]

Some people who were already vulnerable to poverty, hunger, and livelihood loss before the pandemic have had to resort to tragic alternatives including poaching and logging, thereby increasing the degradation of natural ecosystems during the pandemic.[20,21] What's worse, there have been stories all over the news highlighting how crops and live animals have been discarded, wasted, plowed over, or euthanized as a result of supply-chain disruptions.[88,89]

Now imagine a situation in which this food had gone to vulnerable individuals instead of being wasted. With many people out of work or unable to obtain a daily wage during the pandemic, far too many resources and far too much food have been destroyed.[90] We are witnessing these things happening around the world *now*, and they are being intensified by the pandemic. These are things we can learn from and things that we can mitigate with appropriate action, effective governance, and foresight.

Given that diets heavy in animal products contribute to the development of chronic diseases, climate change, environmental degradation, and the outbreak and spread of new infectious diseases, it is time to change dietary recommendations (see Chapter 2). We must fix our food systems and dietary habits now to make them more compatible with the future we want. For more guidance on how, please see Part 2 (Chapters 8 and 9).

2 POLITICS AND DIETARY GUIDELINES
Two Major Problems

> Climate change is the greatest crisis humankind has ever faced, and it is a crisis that will always be simultaneously addressed together and faced alone. We cannot keep the kinds of meals we have known and also keep the planet we have known. We must either let some eating habits go or let the planet go.
> —Jonathan Safran Foer in *We Are the Weather*

Since Chapter 1 covered some of the ways in which agricultural practices contribute to climate change and environmental degradation, it seems only fitting that we also discuss some of the politics that go into creating, producing, and disseminating dietary guidelines, which "are designed to guide individuals to healthier food choices."[§§§]

While there are many versions of dietary guidelines from around the world I could have chosen to discuss, I chose the dietary guidelines that come from the US Department of Agriculture (USDA) for two reasons: firstly, I live in the United States and I am well-informed about these dietary guidelines; and secondly, the US dietary guidelines serve as a particularly "great" example of the "politics" that go into creating dietary guidelines, which undoubtedly occurs in other countries around the world as well.

Despite the knowledge and scientific understanding that certain agricultural practices contribute to climate change and environmental and ecosystem degradation; it is shocking that the US dietary guidelines

[§§§] US Department of Agriculture's Dietary Guidelines for Americans.

would willfully ignore these dangers and actually *encourage* the consumption of foods that produce adverse effects on our own health (by increasing the risk for developing noncommunicable diseases) and on the environment (through toxic agricultural runoff, greenhouse gas [GHG] emissions, deforestation, and overfishing).[91]

When consumers try to follow the dietary guidelines as they are written – because they expect them to represent a healthy diet – they are unknowingly contributing to environmental degradation and potentially harming their own health at the same time.

The majority of chronic diseases such as hypertension, diabetes, heart disease, kidney disease, and cancer – diseases that cost a lot of money and health care resources to treat – are associated with the foods we eat and the foods we are *encouraged* to eat as part of a "healthy diet." The United States has the world's most expensive health care system in the world, but it is designed to treat rather than prevent chronic diseases.[92]

It is ironic that many people will not think twice about undergoing major heart surgery to replace a clogged artery, yet when these same people are asked to change their diet to avoid needing this lifesaving heart surgery in the first place, they look at you as though you have two heads. I know this because I've been on the receiving end of many of these disbelieving looks.

And while many people will make the effort to shift their diets to emulate what is written in the dietary guidelines, believing that these guidelines will promote better health, it is very discouraging to know that too many of the "recommended foods" in these guidelines are neither particularly healthy nor particularly sustainable.

There appears to be little urgency to change these recommendations, or even to debate them. Too many politicians deride what nutrition science says about the health risks posed by some of these foods or deny the risks some of these foods pose to food security and environmental sustainability.

While most Americans do believe human-induced (anthropogenic) climate change is real, many do not believe it will negatively affect their lives.**** Additionally, many Americans are not aware that their

---

**** Anthropogenic climate change refers to the "climate-warming trends" that are largely the result of human activities (https://climate.nasa.gov/scientific-consensus). Humans have caused most of the changes observed in climate and weather patterns.

dietary choices affect the environment (though many are aware that their dietary choices affect their health).[93] Some of this naiveté is related to the fact that this information is not widely disseminated in the news, and some of it is related to the fact that dietary recommendations simply do *not* address the issue in a meaningful way.

Each year, humans consume Earth's resources more quickly than they did in the year before due to population growth, the foods we eat, and a wasteful consumer culture.[94] In just the past 100 years, the human population has increased more than fivefold from 1.5 billion to today's 7.8 billion. By 2100, there may be more than 11 billion people on this planet.[95] Current dietary guidelines and agricultural policies promote this overconsumption and are irresponsible in that they encourage eating patterns that can exacerbate climate change, are not sustainable, and may negatively affect food and livelihood security for all of the reasons discussed in Chapter 1.[96]

The connections among dietary guidelines, agricultural policies, environmental degradation, and chronic disease demonstrate that there is a need to change dietary patterns and paradigms sooner rather than later.

We have only one planet. We must take care of it and its resources to guarantee food security, health, and environmental sustainability into the future.

The Disconnect between Dietary Recommendations and Environmental Repercussions

Three-quarters of American adults believe dietary guidelines should include recommendations concerning the environment and sustainability.[97] To be fair, the nutrition science experts who comprised the 2015–2020 Dietary Guidelines Advisory Committee (DGAC) recommended that sustainability and food security issues *be incorporated* into the Dietary Guidelines for Americans (DGAs), stating that "a diet high in plant-based foods is ... associated with less environmental impact than the current U.S. diet."[98],††††

Instead of incorporating sustainability recommendations into the DGAs, the DGAs that were *actually published* describe dietary patterns that are not sustainable. The DGAs recommend moderate

††††   The 2020–2025 DGAs have not yet been published or made available.

intake of foods that increase environmental degradation, overfishing, deforestation, and the incidence of chronic disease. Specifically, the DGAs recommend "moderate amounts of seafood," "dairy products, including milk, yogurt, and cheese," and "lean meats, poultry, and eggs."[98]

However, as addressed in Chapter 1, many of these foods perpetuate environmental degradation or climate change in one way or another.[99],‡‡‡‡

Physicians take the Hippocratic Oath, stating, "First, do no harm." Yet I believe certain recommendations found in the DGAs cross ethical boundaries, potentially resulting in detrimental "health, environmental, social, and economic" consequences that *do* cause harm.[100] Many countries around the world use terminology or intake recommendations that are similar to those found in the DGAs, indicating that many of these recommendations, while not universal, are widespread.[101] Here are some examples.

Fish

The American Heart Association (AHA), the DGAs, many European countries, and many other countries around the world recommend at least two servings of fatty fish per week.[102,103,104] The AHA *also* recommends omega-3 fatty acid pills (eicosapentaenoic acid and docosahexaenoic acid), which are most often derived from fish oils, for individuals with coronary heart disease.[102]

While very high doses of fish oil may benefit individuals with overt heart disease, for the general population, the evidence suggests that a whole-food, plant-based diet – high in fiber, monounsaturated fats (olive oil, peanut oil, sesame oil, avocado), and plant-based omega-3 fatty acids (chia, flax, walnut) – is more effective at preventing and mitigating heart disease than is adding fish and fish oil (to a not-so-great diet).[105,106] Yet most people who take fish oil are doing so without the express recommendation of their physician, and they take it because of its purported benefits (for more on this, see Recipe 3).[107]

---

‡‡‡‡ These food products contribute to environmental degradation or climate change by way of their GHG emissions (methane from dairy and beef cows), deforestation and land-use change (dairy and beef cows, feed for egg-laying hens/broiler chickens, and cows), agricultural runoff from farms, and environmentally destructive fishing methods (this will be discussed in Chapter 3).

This is problematic, as large proportions of the world's fish stocks are overexploited or fully exploited. Roughly 90 percent of large, sexually mature fish and more than 97 percent of southern bluefin tuna have already been extracted from the oceans.[108,109,110,111] These severe reductions in fisheries biomass have serious implications for global food security, as nearly 3 billion people depend on fish as part of their daily diet and/or livelihoods.[112]

Fish make up one-sixth of the human population's intake of animal proteins. Nearly 40 percent of all people eat fish to meet at least 20 percent of their protein needs, and over half (55 percent) eat fish to meet at least 15 percent of their protein needs.[113] Populations in coastal developing countries engage in subsistence fishing, meaning they take only what they directly consume. These people face many threats from depletion of this indispensable food source due to illegal fishing by fleets from other countries or their own governments selling off fishing rights to wealthier countries.[114,115]

Industrialized countries are not immune to overfishing or fisheries depletion either. In the 1990s, cod populations collapsed off the Atlantic coast of Canada due to "[an] inability to control [catch rates]," low survivability from birth to three years of age, and significant reductions in sexually mature individuals.[116] Today, both direct and indirect threats to fisheries affect their sustainability and viability. Dietary guidelines that recommend fish and fish oil consumption at their current levels put these fisheries at further risk, particularly when there are healthier and more effective alternatives available.

Direct threats to fisheries include industrial fishing methods, namely the use of long lines and nets. Bottom-trawling, purse seine,§§§§ and gill nets***** all destroy coral reefs and ocean-bottom habitats (even those in deep water). These methods also capture both the intended target species and a myriad of unintended species, known as by-catch or by-kill. Depending on the target species, the fishing method, and location of the catch, by-kill can range between 7 and 98 percent, killing hundreds of thousands of marine animals every year.[117] This is not sustainable and may have far-reaching consequences for food security in the future.

---

§§§§ Purse seine nets are nets that cinch at the top and bottom, like a purse, to capture fish.
***** Gill nets are like a curtain of netting that hangs down from the surface of the water.

Indirect threats to fisheries include climate change, warming oceans, acidification, and various pollutants (more on this in Chapter 3).[118] Oceans account for more than 90 percent of Earth's habitable space (for both animals and plants). Oceans also delay the worst effects of climate change by slowing down the rate of warming. Because of their size and sheer volume, oceans are a huge "heat sink," meaning they absorb a lot of heat and moderate Earth's temperature. But even with this "delaying" effect, Earth is warming as a result of anthropogenic GHG emissions – including all of those previously discussed.[119]

Oceans are also a gigantic carbon sink. Tiny oceanic organisms produce more than half of the oxygen we breathe and absorb or sink more than half of the carbon we emit.[108] An increase in dissolved carbon in the oceans (leading to acidification) and warmer ocean temperatures harm coral reefs and other organisms that depend on calcified structures, reduce krill and fisheries fecundity, and alter entire food chains and ecosystems. As well as being damaging to life in the oceans, these changes affect resources that humans also depend on.[119,120,121] It's not just about food. Increasingly, many drugs are derived from ocean animals and/or plants.[†††††] If those animals and plants no longer exist – or are reduced significantly in number – then think of all the future drugs or therapeutics we may not have access to.

Plastic – an increasingly pervasive threat – is harmful to marine animals' health and our own. Nearly 8 million metric tons of plastic debris enter the oceans every year.[122] When they ingest it, fish and other marine animals store plastic and other chemical by-products in their fats, where they bioaccumulate up the food chain to larger animals, and finally to humans. In Indonesia, anthropogenic debris (primarily plastic) was found in 28 percent of individual fish's digestive tracts and in 55 percent of all edible fish species. In the United States, anthropogenic debris (primarily plastic-based fibers) was found in 25 percent of individual fish's digestive tracts, in 67 percent of all edible fish species, and in 33 percent of individual shellfish's digestive tracts.[123]

By 2050, there may be more plastic (by weight) in the oceans than fish! There will be increased exposure to toxic chemicals, carcinogens, and other endocrine-disrupting compounds, with the potential for

††††† This includes the COVID-19 vaccine, which requires the blood of horseshoe crabs (www.nationalgeographic.com/animals/2020/07/covid-vaccine-needs-horseshoe-crab-blood).

public health consequences (for more on this, see Chapter 4 and Table 4.1).[124]

These direct and indirect threats pose a dilemma for consumers, faced as they are with the seemingly contradictory medical and dietary guidelines to eat seafood. These threats also present a dilemma for policymakers tasked with meeting the nutritional needs of their populations while also caring for Earth's precious and under-threat resources. It is time to look to alternatives *now*, before it is too late. In Part 2 of this book, I present several potential – and sustainable – ways forward.

Agriculture

The domestication of plants and animals – but primarily plants – as part of the Agricultural Revolution allowed human populations to grow rapidly. Modern agricultural methods have ensured adequate energy availability (calories and protein) throughout the world.[125,126,‡‡‡‡‡] Unfortunately, some modern large-scale agricultural methods – especially those related to animal agriculture – produce negative externalities that contribute to GHG emissions and habitat degradation, to say nothing of the well-being of the farmed animals themselves.

As noted in Chapter 1, agriculture produces at least 25 percent of GHGs globally (but likely more). Livestock produce more than half of that amount – at least 14.5 percent of all GHG emissions globally (7.1 gigatons/year), which is more than the entire global transportation sector. It is estimated that up to 50 billion head of livestock are raised each year around the world.[33] The Paris Climate Agreement aimed to lower GHG emissions to a level that will prevent the global average temperature from rising more than 2°C (3.6°F). Yet the Agreement focuses primarily on reducing fossil fuel use. Unless we change our agricultural and dietary practices, GHG emissions will increase from the agriculture sector due to human population (and companion animal) growth and the increased desire for animal products.[127,128]

Simply put, an increase in animal agriculture will threaten global food security and health.

In 2016, one-third of all cereals grown and 42 percent of all arable land went to feed agricultural animals. It takes several hundred

‡‡‡‡‡ Unfortunately, though, these are not well distributed, as is addressed in Chapter 1.

pounds of grain and several thousand gallons of water to feed just one animal. As global demand for animal products increases, deforestation will continue at ever-increasing rates either to create feedlots or to increase grain production. Yet the removal of forests eliminates carbon sinks,§§§§§ exacerbates GHG emissions, increases soil erosion, and strains land and water resources in many regions that are already resource poor, further reducing resilience to hunger and food insecurity.[82,126,129]

One solution with great potential for significant positive effects is to transition to a more plant-based diet and to move away from high levels of animal foods consumption – especially in developed countries, where animal foods consumption is very high. Another benefit is that in addition to reducing GHG emissions and conserving water resources, there is good evidence that plant-based diets reduce the risk of chronic disease and can sustainably feed a growing world population.[34,130,131,132]

It is true that the DGAs mention plant-based proteins. However, they are near the end of the list of recommended protein-rich foods. Instead, the DGAs – perhaps influenced by agricultural lobby interests – encourage greater consumption of "lean meats," dairy, and eggs, and they also direct consumers to less sustainable, less healthy food choices, including eggs, milk, yogurt, chicken, fish, and occasional meat products.[133] The animal agriculture industry does not want to lose consumers and will go to great lengths to prevent that from happening. However, there is hope, as recent trends in food purchasing show an increased desire for plant-based meat and dairy "analogs," especially among younger generations and those who are worried about the environment.[134] This trend is good news, but it is worrying that the DGAs are not the ones leading the charge and recommending healthier plant-based options. Instead, the DGAs seem to bend to the will of the agriculture industry.

Additionally, there are some meat-producing companies that are suing companies that produce nonanimal meat analogs for calling their products "meat." That's not great. But it is worth noting that there are also some meat- and dairy-producing companies that *are* investing

---

§§§§§ A carbon sink is a natural or artificial reservoir that absorbs and stores the atmosphere's carbon through physical and biological mechanisms. The ocean, plants, trees, and soil are some examples of carbon sinks (https://ocean-climate.org).

in companies that produce plant-based meat and dairy analogs, demonstrating a commitment to preventing the worst effects of climate change from the food industry. Or – more cynically – perhaps they are demonstrating a desire to get into where they believe the future of "meat" and "dairy" will be.[135]

Evidence suggests that diet is far "more effective than technological mitigation options for avoiding climate change and ensuring access to safe and affordable food for an increasing global population." Diet is equally effective at reducing the incidence and prevalence of noncommunicable nutrition-related chronic diseases[34] – all good things!

Food Waste

As mentioned in Chapter 1, more than one-third of all foods are wasted. This produces 4.4 gigatons of carbon emissions each year (7–8 percent of total global emissions).[36,37] Food waste occurs when produce (and sometimes animals) gets discarded either on farms – never making it to market – due to spoilage along the supply chain or in the home by consumers. Just imagine how many *more* people could be fed if this food were not discarded.******

Worldwide, agriculture uses over 70 percent of all the fresh water that is withdrawn. In the least developed countries, this figure rises to more than 90 percent of the fresh water that is withdrawn.[136] When food is wasted, critical land and water resources are squandered.[137]

Population growth and the demand for more animal products underscore the need to significantly reduce food waste to ensure an adequate and safe food supply while also moderating future environmental impacts.[34,138] While not technically a dietary guideline, addressing the issue of food waste seems invaluable.

Recommendations

There is good evidence that we can mitigate chronic disease, food insecurity, environmental degradation, and agricultural emissions

---

****** For additional information, see the USDA's website on food waste: www.usda.gov/foodwaste/faqs.

through diet. Shifting health and agricultural policies to support greater intakes of plant-based diets can reduce global morbidity and mortality by 6–10 percent and agricultural GHG emissions by between 29 and 70 percent.[34] This is great news!

However, there are barriers to accomplishing this shift, which include political will, special interest groups, lobbyists, and corporations that support the "business-as-usual" model – intensive animal agriculture – and have the resources and the influence to convince governments and "health organizations" to support their goals.[133]

As a case in point, in the United States, agricultural policies actually *subsidize* corn, sugar, wheat, alfalfa, and the use of concentrated animal feeding operations in animal farming. These subsidies often cause plant-based foods to be diverted away from human consumption so that they can be fed to animals (or sometimes they are used to produce biofuels), supporting unsustainable and deleterious dietary patterns. These subsidies, which ironically are paid for by our hard-earned taxes, provide extreme discounts on the true costs of these foods to our health, the environment, and food security. But believe me when I tell you that we pay in other ways – in the form of high rates of chronic disease and health care expenses related to treating those chronic diseases.

In another example of misleading recommendations for what constitutes a healthy diet, we have witnessed the US Defense Production Act be put to use to encourage meat processing and packing plants to reopen amid the COVID-19 pandemic lockdowns. The reason? Because meat is a "scarce and critical material essential to national defense."[139] This is their reason, not mine, and this is simply not the case. Adding insult to injury, invoking this Act removed the meat processing and packing companies' liability and legal exposure should any of their employees get sick from the virus.

The USDA has time and time again allowed certain food industry groups and their well-funded lobbyists to undermine the science and recommendations of the independent DGAC. This ultimately leads to the dissemination of dietary guidelines that promote and essentially recommend foods that harm our health, exacerbate chronic disease, and harm the environment.[140]

Since the advent of agriculture, humans have destroyed 83 percent of all wild mammals and half of all plants, with most of this occurring since the Industrial Revolution.[84] These losses, in part, are due to *us* raising billions of animals – from just a handful of domesticated species for human consumption – that destroy the habitats of wild animals.[141] Earth is now on target to lose more than two-thirds of species within the next decade as a result of human dietary choices, poaching, ongoing habitat loss, and climate change.[142,143]

By 2050, 50 percent of all coral reefs around the world could be lost permanently due to persistently warmer (and more acidic) oceans. While reefs account for only 1 percent of the world's oceanic habitat, they support nearly one-quarter of all oceanic life through the nutrients they produce and the shelter they provide.[144] This is an enormous loss we are talking about.

So much loss in biological diversity – as a result of our actions – reduces the resilience of the entire system.

It is expected that greenhouse gas emissions will decrease from the energy and transportation sectors over the next few decades due to the development and use of more environmentally friendly technologies. But we must not forget the GHG emissions produced from the agriculture sector for at least a couple of reasons:

(1) There will be more people to feed; therefore, food production will need to expand, likely leading to increased deforestation.
(2) Dietary choices, especially in developing countries, are shifting toward higher intakes of animal products. This shift in food choices will increase the amount of GHG emissions produced from this sector immensely.

Sadly, most people have not been made aware of the relationship between diet and the environment, mainly because it is not widely communicated in the media or even in national dietary recommendations. Concerns surrounding climate change have been singularly focused on fossil fuels, to the detriment to our health and the environment. Yet, "We will never address climate change, never save our home until we acknowledge that our planet [has become] an animal farm," one that is causing far too much harm.[84]

There is conflict between government-decreed dietary guidelines and dietary patterns that are known to benefit human health, food security, and the environment. Dietary guidelines often portray themselves as healthy, but they recommend too many animal-based foods. In contrast, plant-based and Mediterranean-style diets, both of which are low to minimal in meat and dairy products, are far better for you and the planet.

This chapter has attempted to highlight some of the dietary and environmental issues inherent in our political systems and nutrition guidelines – which are *supposed* to make us healthier – and calls us all to action to better understand and disseminate the urgency of these issues.

Humans are adept at finding solutions to problems. However, humans also seem to be all too keen to ignore problems that we find inconvenient, choosing instead to "live for the moment," and often at our own peril. We tend to believe that – when it comes to climate change – our individual role doesn't matter or make much of a difference, and while that may be true when it comes to broad-sweeping environmental policies or regulations, it is not the case for our diets.

When we cannot get governments to change policy for us all, we as individuals can and must change the path we are on, one person at a time.

We can help reduce chronic disease and promote health, food security, and sustainability all at the same time by shifting to more healthy and sustainable options.[29,34] Doesn't that sound worthwhile?

The more individuals who implement these changes and demand healthier options for themselves and the planet, the more agriculture and business will feel forced to change and the more governments will feel forced to recognize the need for change. But this will only happen when market forces kick in, from the ground up, to create change.

We urgently need better policies, dietary guidelines, and communications that inspire and foster healthier, more sustainable consumption habits for our populations, but sometimes this inspiration must come from us.

If we continue to ignore these problems or allow political and business interests to overshadow them and do nothing, we simply imperil human health, the environment, and our (and other life-forms') ability to live here. We must act now. Part 2 of this book will provide some tangible recommendations and ideas – "recipes" – on how we can do this.

In case you wanted to know more about how dietary guidelines are created in the United States, take a look at Figure 2.1.

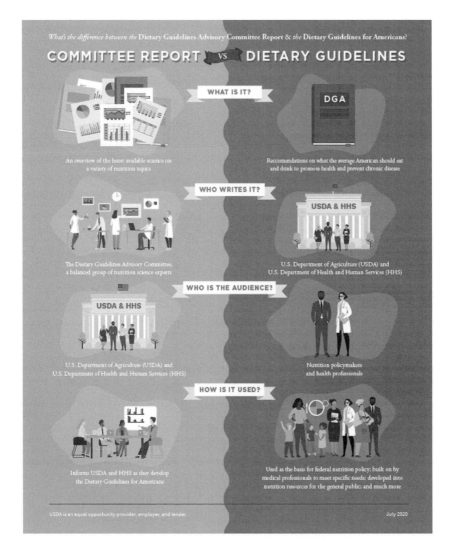

Figure 2.1 US Department of Agriculture graphic explaining the process of producing dietary guidelines.
Image credit: US Department of Agriculture

3 OVERFISHING
There Is No Such Thing as Sustainable Fish

Protect the natural systems as if your life depends on it because it does!
Everyone, everywhere is inextricably connected to and utterly
dependent upon the existence of the sea.
—Sylvia Earle (aka Her Deepness)

Since I spent part of the last chapter addressing the use of fish
and fish oils in dietary recommendations, I feel it is only appropriate to
provide a more in-depth chapter further addressing these issues and why
they are not so sustainable.

Right now, we are in the middle of a big experiment. We are
experimenting with our oceans, our fisheries, and our climate. When we
walk into our local grocery store or fish market, we walk into one of
the biggest lies of all. We see stacks of fish and piles of shellfish at the
seafood counter. We see dozens of live lobsters walking around in their
tanks with little rubber bands around their claws. We see beautiful flaky
white fish, thick fatty orange salmon, and deep red tuna.

We think nothing of it because the fish are always there. It does
not matter what day of the week, the month, or the year it is, the fish
and seafood are always there, ready for us to purchase, take home,
cook, and eat.

But this may not always be the case.

Read the book *The End of the Line* or watch the movie of the
same title – both detail the plight of the world's oceans and their
wildlife.[115] The United Nations Environmental Programme (UNEP)
and the Food and Agriculture Organization of the United Nations

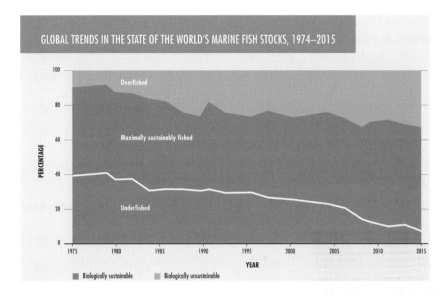

Figure 3.1 Global trends in the state of the world's marine fish stocks, 1974–2015
Image credit: Food and Agriculture Organization of the United Nations (2018) –
State of World Fisheries

(FAO) report that oceanic ecosystems are increasingly at risk of over-fishing, as well as illegal, unreported, or unregulated fishing.[111,145] Humans are depleting the oceans of their fish – their life.

We have fished more than 33 percent – one-third – of the world's fish stocks at unsustainable levels, and we have fully fished nearly 60 percent of the world's fish stocks at maximally sustainable levels, where additional or increased catches will render them unsustainable.[††††††] In sum, more than 90 percent of all fish stocks are either already at their maximum production yield or else are overfished.[111]

The proportions of overfished and maximally fished stocks have grown since 1974 and continue to grow every year (see Figure 3.1). At the same time, fish stocks that are underfished have decreased from 1975 to 2015, such that as of 2015 only 7 percent of the world's fish stocks were reportedly underfished. And we are moving down the seafood chain. Fish that are caught today are often younger and smaller than their predecessors were. This means (depending on the species) that

---

[††††††] Unsustainable fishing/overfishing = capturing or harvesting fish at a rate that sees declining populations of that species over time. Maximum sustainable yield (maximally sustainable level) = the highest possible amount, level, or yield of a resource or species stock that can be routinely caught (annual catch) and sustained over time.

they may not have had the chance to reproduce yet, further endangering the long-term viability of those species.[111]

Just recently, a number of "table species"[‡‡‡‡‡‡] have been declared critically endangered and threatened with extinction by the International Union for Conservation of Nature (IUCN).[§§§§§§] Metaphorically speaking, eating certain species of fish is somewhat similar to the idea of eating a rhinoceros, an elephant, an orangutan, or a cheetah.

The Intergovernmental Panel on Climate Change (IPCC) and the United Nations Framework Convention on Climate Change (UNFCCC) expect climate change (and ocean acidification) to create several problems that will influence the health of the oceans and fisheries. One example of these likely problems is ocean acidification. As more carbon is added to the atmosphere and absorbed by the oceans, the pH of the oceans drops. This makes the oceans more acidic than they were previously. Acidification can negatively affect the life cycles of certain shellfish and corals, especially those that depend on calcium-rich structures. The more acidic the oceans – as compared to their baseline – the more challenging it is for these creatures, *especially corals*, to build their shells and skeletal structures in order to survive.[146]

When ocean temperatures increase, this also negatively affects corals and coral reefs. Corals thrive within a very narrow temperature range and are slow to adapt to changes. When they cannot adapt fast enough to rising temperatures, they bleach and may eventually die. A reef cannot simply move to a cooler, more habitable part of the ocean. The bleaching and death of wide swaths of coral reefs as a result of increasing temperatures have knock-on effects for the species that depend on them for food and shelter, as well as for the humans who depend on these reefs and the life they support for food.[147]

Oceanic "dead zones" are areas in the ocean where there is an insufficient amount of oxygen for fish or other marine life to survive. Dead zones are typically found at the end points or mouths of rivers, where nitrogen- or phosphorus-rich effluent drains out into the oceans. The most common sources of nitrogen- or phosphorus-rich effluent are

---

[‡‡‡‡‡‡] Table species are species that are commonly consumed (e.g. salmon, tuna, trout, and grouper).
[§§§§§§] The IUCN is an international organization working in the field of nature conservation and the sustainable use of natural resources (www.iucn.org).

animal waste products from concentrated animal feeding operations and/or excess fertilizer from farms that run off into rivers. Dead zones are growing in both size and number around the world as these nutrient-rich wastewaters lead to the overgrowth of algae, remove oxygen from the water, and suffocate animal or plankton species that depend on drawing oxygen from water.[148]

Increased temperatures, acidification, and the spread of dead zones are just some of the insults occurring in the world's oceans. If we carry on as we are, primary production of fish from the oceans is expected to decline by 6 percent by 2100, and by as much as 11 percent in tropical zones.[111] Yet there will be 2 billion additional people on Earth by 2100, or even more. Foods derived from the oceans will be needed to help feed all of these additional people.

In 1961, when the world's population was roughly 3.1 billion people, per-capita intake of fish or fish products was around 20 pounds (9 kg) per person per year. Today, with a global population of 7.8 billion people, per-capita intake of fish is nearly 45 pounds (20.5 kg) per person per year. There are more than twice as many people, each eating twice as much fish. World fish production, consumption, and trade are expected to increase, yet the growth rate of fish and fish products has been slowing over time (see Figure 3.1).[*******] How do we square the likely decrease in fish production from the oceans with the expected increase in the consumption of fish and fish products around the world due to an increased population? Clearly, we can't.

And while we have been led to believe that consuming farmed fish is a more viable, sustainable, and ecological choice, and many special interest groups are bent on convincing us of this as fish farming has been scaling up, in some instances farmed fish may be even worse for the environment and fisheries sustainability than wild-caught fish.

Farmed fish are more likely to harbor pests and diseases than wild fish, and they may also require specialized dyes or other chemicals in their feed. Furthermore, depending on the species, pound for pound, certain farmed fish may consume greater amounts of wild-caught fish feed than they produce in farmed fish flesh for us to eat, which is ecologically unsustainable.[149]

[*******] The State of World Fisheries and Aquaculture 2018 (www.fao.org/state-of-fisheries-aquaculture).

For example, it can take up to 5 pounds (4.9 pounds to be precise, or 2.2 kg) of small wild-caught fish such as herring, menhaden, or anchovies to feed and grow 1 pound (450 grams) of farmed salmon. Farmed trout require just over 3 pounds of wild-caught fish to produce 1 pound of farmed flesh. Even farmed shrimp require a little more than 1 pound of wild-caught fish to produce 1 pound of edible farmed flesh. These ratios are quite inefficient from a resource perspective.[149]

Not only that, but the levels of healthy omega-3 fatty acids found in farmed salmon and trout – pound for pound – are lower than in smaller wild-caught fish such as anchovies or menhaden. The lower omega-3 levels found in farmed fish are primarily because farmed fish are fed food sources – in addition to small wild-caught fish – that do not contain omega-3 fatty acids.[150] And while there are some species of farmed fish that require lower amounts of wild-caught fish feed, such as tilapia, catfish, and carp, these farmed fish are typically less "desirable" to eat, as they contain significantly less omega-3 fats than do salmon, trout, herring, or anchovies.[149]

As it stands, depending on the type of fish and methods used, fish farming is not the best or most efficient use of a global resource that is already on the decline.

To further exacerbate these issues, medical institutions and nutrition academies frequently recommend fish oil supplements for better health. Yet this, too, is unsustainable. Recent forecasts, models, and statistical analyses indicate that in a "business-as-usual" world we could diminish the fish species we typically eat by 2048 to the point at which they are no longer viable options to eat as food, as it would be too costly or inefficient to catch them.[151]

We are simply taking fish out of the oceans faster than they can reproduce, and we are consuming fish (directly and indirectly) at rates that are too high to be ecologically sustainable. This could mean too few tuna, Chilean sea bass, grouper, or other fish species available for commercial fishing, meaning that many people around the world who depend on fish for their livelihoods or for their nutritional security could be at risk of poverty or hunger.

Nearly one-fifth of the world's population – roughly 1.2 billion people – depend on the oceans for their main source of protein and as their primary livelihood.[152]

According to a report by the World Bank in collaboration with the FAO, the University of Arkansas at Pine Bluff, and the

International Food Policy Research Institute (IFPRI), "Supplying fish sustainably – producing it without depleting productive natural resources and without damaging the precious aquatic environment – is a huge challenge." This report further states that "we continue to see excessive and irresponsible harvesting in the capture fisheries and in aquaculture."[153]

The global population is expected to grow by another third by 2050, nearing 9.7 billion people. Yet some fish stocks are already on the verge of collapse.[154] And while aquaculture and fish farming are growing and currently provide almost half (47 percent) of all fish produced in the world, there are problems associated with aquaculture aside from just its (feed) inefficiency. As an example, another concern with farmed fish is that some contain a greater burden of natural and human-made toxic substances such as antibiotics, pesticides, and persistent organic pollutants than wild fish. This is a result of what they are fed and their living conditions, potentially posing health problems to those who regularly eat farmed fish.[155] Additionally, farmed fish produce a lot of waste products and increase the potential for disease outbreaks that can spread to wild populations who may share the same waterways.[111,156]

Clearly, there is much room for fish farming and aquaculture methods to improve, be safer, and become more efficient in the future with advancing technologies.[157] If we remove too much from the oceans, destroy habitats, and exacerbate ocean (and global) warming, we increase the risk that there will be a global food crisis – one that aquaculture alone will not be able to fix. It is also worth noting that food crises may increase the risk of regional or global security crises, as demonstrated by the fact that humans have been fighting over resources for centuries.[158] The oceans and their sea life provide services and resources that millions of people depend on and would more than likely be willing fight for.[159]

There are several organizations that advocate for consuming sustainably caught or farmed fish, yet the evidence that this is possible – to catch or farm fish sustainably *and* restore natural resources seems – rather sparse thus far. So what can you do in the meantime?

(1) Please do not order or eat bluefin tuna. Atlantic bluefin tuna are an endangered species and southern bluefin tuna are a critically endangered species. Pacific bluefin tuna are a species that is

vulnerable to extinction. It would not be a stretch to say that eating bluefin tuna would be similar to ordering elephant or rhinoceros in a restaurant.

(2) If you feel you must eat seafood, there are more sustainable options lower down the food chain. Smaller fish such as anchovies or herring tend to reproduce in significantly higher numbers and rates and at younger ages. Discuss with your grocer or fishmonger from where their fish is sourced. Obtain as much information as possible to ensure that what you are eating is not an endangered species simply renamed for marketing purposes.

(3) Skip the fish oil. There is simply no good evidence that fish oil provides any major health benefits. It is extremely expensive for what you get and it is not sustainable. Additionally, we can get omega-3 fatty acids from other sources, including algal oils, flaxseed, seeds, and nuts (there will be much more on this in Recipe 2 in Part 2 of this book).

4 PLASTIC
It's What's for Dinner

Only when the last tree has died and the last river been poisoned and
the last fish been caught will we realize we cannot eat money
[or plastic].
—Cree First Nations Prophecy

The Prevalence of Plastic

How many plastic items can you name? Ten? Twenty?

Did you know that in the United States we each use approximately 150 individual plastic bottles each year, and yet only around 35 of those bottles will end up in the recycling bin?[160,161,†††††††] In Europe, the recycling figure is a bit better. Mainland Europeans use roughly 103 individual plastic bottles each year and recycle an average of 31 of those.[162] Meanwhile, in the UK, nearly 195 plastic bottles are used each year and roughly 112 of those get recycled (good job on recycling, UK!).[163]

In addition to plastic bottles, we each use over 300 single-use plastic bags every year in the United States (100 billion for the entire country).[164] Yet we only recycle about 5 percent of those bags.[165] Around the world, 4 trillion plastic bags are used every year (that's an

††††††† This includes plastic bottles of water and/or other individual beverages such as plastic bottles of soda or pop.

astounding 513 per person), but only 1 percent of those bags are returned for recycling.[161]

This is a huge problem. Although plastic seems to have become as present in our lives as the air we breathe – it's convenient, it's light, it's cheap, and you can throw it away if you want to – it is now invading our oceans, it is being ingested by fish and sea life, and it is even making its way into our own bodies. This is not good for the planet or for our own health.

Modern plastic was first created in 1907 by Leo Hendrik Baekeland. Today's major thermoplastics – including polyvinyl chloride (PVC), low-density polyethylene, polystyrene, and polymethyl methacrylate – were developed and commercialized in the 1930s and 1940s.[164] World War II saw a significant increase in demand for thermoplastic materials as replacements for some of the raw and natural materials (such as natural rubber) that were difficult to obtain during the war.[166] During this period, military operations monopolized the use of plastics.[167]

Between 1939 and 1945, plastic production increased fourfold, from 106,000 tons per year to 409,000 tons per year.[164] After the end of World War II, plastic became a preferred and oft-used material, primarily due to its versatility and usefulness for many applications. By 1950, global plastic production reached 1.7 million tons per year.

What were the reasons for this rapid increase?

Plastic was cheap, convenient, easy to make, considered disposable, and virtually indestructible. It still is! Plastic can be molded into almost any shape or form and has successfully replaced materials like glass and metal in many products.

Now, take a moment and read the following quote from a 1950s commercial: "This is Tupperware ... Tupperware comes in all shapes, all sizes. Since Tupperware won't leak, you can freeze it, stack it, any which way ... Just look in the yellow pages under housewares or plastic."[‡‡‡‡‡‡‡] This commercial, like many others, was commonly seen in the 1950s when plastics first entered our lives in a mainstream way.

Today, global plastic production has expanded so much that it nears 380 million tons each year. This is a 223-fold increase since 1950.[168,169] During this same time period, the global population has

‡‡‡‡‡‡‡ https://youtu.be/1u9kEnJZX2E

only increased threefold, from around 2.5 billion people to roughly 7.8 billion people.[170] So where is all this plastic going?

Plastics are used in all types of consumer products: bottles, bags, telephones, clothing, flooring, and car parts. About 10 percent of the weight of a typical car – about 336 pounds (153 kg) – is plastic. That's roughly 50 percent by volume. Plastic is typically used in the bumpers, door panels, and engine components.[165]

Plastic nylon is used in fishing lines and nets, and it is highly durable.[171] Fishing lines and nets are discarded at very high rates, making them some of the most common plastic discards found in the ocean. This is also why discarded or lost fishing gear is so very dangerous – they do not disintegrate. Instead, they remain in the ocean "ghost fishing," killing hundreds of thousands, if not millions, of animals every year.

Look around you right now and see how many objects are made of plastic. You may be surprised. Today, plastic accounts for more than 30 percent of all packaging sales. It really is everywhere.

How Is Plastic Made and Why Is It Such a Problem?

Plastic is made from oil, natural gas, and other petroleum-derived chemicals. Today, roughly 4 percent of the world's petroleum is used to make plastic materials, and another 4 percent is used to power the plastic manufacturing processes.[172]

The thing about plastics is that they are virtually indestructible. They do not biodegrade in the way paper products can and they do not disintegrate. Plastics persist in the environment for thousands of years.[173] The more plastics we make, the more plastics end up in our landfills and, inevitably, in our oceans.

Each year, over 8 million metric tonnes (8.8 million tons) of plastic waste end up in our oceans. Most often, this plastic waste comes from discarded fishing gear and packaging that is meant to be used only once and then disposed of.[174] Figure 4.1 demonstrates how plastics enter the oceans.§§§§§§§

§§§§§§§ A metric tonne (aka metric ton) is a unit of weight that is equal to 1,000 kg, or roughly 2,200 pounds. A "ton," on the other hand, is an English unit of measurement equal to 2,000 pounds, or roughly 907 kg.

The pathway by which plastic enters the world's oceans

Estimates of global plastics entering the oceans from land-based sources in 2010 based on the pathway from primary production through to marine plastic inputs.

**Global primary plastic production:
270 million tonnes per year**

**Global plastic waste:
275 million tonnes per year**
It can exceed primary production in a given year since it can incorporate production from previous years.

**Coastal plastic waste:
99.5 million tonnes per**
This is the total of plastic waste generated by all populations within 50 kilometres of a coastline (therefore at risk of entering the ocean).

**Mismanaged coastal plastic
waste: 31.9 million tonnes per year**
This is the annual sum of inadequately managed and littered plastic waste from coastal populations. Inadequately managed waste is that which is stored in open or insecure landfills (and therefore at risk of leakage or loss).

**Plastic inputs to the oceans:
8 million tonnes per year**

2 billion people living within 50km of coastline

**Plastic in surface waters:
10,000s to 100,000s tonnes**
There is a wide range of estimates of the quantity of plastics in surface waters. It remains unclear where the majority of plastic inputs end up – a large quantity might accumulate at greater depths or on the seafloor.

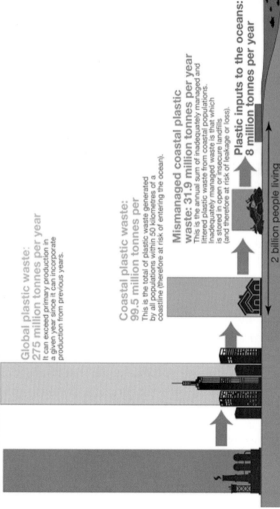

Source: based on Jambeck et al. (2015) and Eriksen et al. (2014). Icon graphics from Noun Project.
Data is based on global estimates from Jambeck et al. (2015) based on plastic waste generation rates, coastal population sizes, and waste management practices by country.
This is a visualization from OurWorldInData.org, where you will find data and research on how the world is changing.

Licensed under CC-BY-SA by the authors.

Figure 4.1 How plastic materials enter the world's oceans

For many of us, plastic bottles and bags are the most common items we throw away. In the United States alone, we use nearly 50 billion plastic bottles each year.[175] Worldwide, that figure is nearly ten times as much, or roughly 480 billion plastic bottles used each year.[161] In addition, roughly 2 million plastic bags are used every minute globally, or nearly 1 trillion per year. This is astounding considering that plastic bags were only introduced into grocery stores in 1977 as an alternative to paper bags.[167] By the mid-1990s, the use of plastic bags grew so much that four out of every five grocery bags used in the United States were made of plastic.[167,176]

The amount of energy and petrochemicals used to produce all of the bottles and bags used in the United States each year would be enough to fuel all 6.4 million cars of Los Angeles for a year.[177] Similarly, the amount of energy and oil used to make just twelve plastic shopping bags could drive your car a mile or more (depending on the car).[178]

In the last few years, several cities in the United States and around the world banned or taxed single-use plastic bags. In 2014, Governor Jerry Brown of California signed a bill banning single-use plastic bags in grocery stores across the state. However, consumers in California could still purchase thicker, "multi-use" plastic bags in grocery stores, and at only ten cents per bag there was little disincentive to using them.[********] Before this ban, Californians were using 120,000 tons (109,000 metric tonnes) of plastic bags every year.[179] Keep in mind that a single plastic bag weighs only five grams, or one-sixth of an ounce.[††††††††] I know this because I weighed one myself. Yet this very small, very lightweight piece of plastic can do a lot of damage.

In response to the California ban, the plastics industry reproached and admonished Governor Brown. Their goal? To get the plastic bag ban overturned.

Fortunately, they failed.[‡‡‡‡‡‡‡‡]

******** With that being said, however, many Californians on their own initiative significantly reduced their use of plastic bags and turned to reusable fabric shopping bags instead.
†††††††† This means that, in California, 23,040,000,000 (23 billion) plastic bags were being used every year on average (that is around 583 bags per person).
‡‡‡‡‡‡‡‡ However, due to the COVID-19 pandemic, grocery stores in the State of California have asked that shoppers not bring in their own reusable bags due to concerns regarding contamination. As a result, the plastic bag ban has been put on hold until further notice and there is currently no cost for using single-use plastic bags in grocery stores. As far as plastic is concerned, this a step in the wrong direction.

Consumers do have the power to *refuse* plastic products and prevent them from entering the environment, but first they need to know that plastic *is* an environmental problem. So far, I have not seen many public service announcements delineating this problem, nor have I observed much messaging on this in the media. With that said, however, I have noticed that this information is starting to make the rounds and more people seem to becoming aware of it.

The Recycling Misnomer

While we theoretically *can* recycle plastic, as the campaigns of yesteryear encouraged us to do (remember "Reduce, Reuse, and Recycle?"), in reality, only about 7 percent of plastics in the United States are actually recycled or recovered. Just think about how many times you have seen someone place a plastic bottle or bag into the rubbish container rather than the recycling one, even if they were right next to each other.

Plastics that are recovered through the "recycling process" cannot really be recycled or repurposed in the same ways that glass or metals can be. Glass, aluminum, and other metals can be fully melted, sterilized, and reformed into new glass, aluminum, or other metal products without degrading. This saves both energy and raw materials.[180,181] Recovered plastics, however, instead of being "recycled" are often "downcycled" into other products such as park benches, doormats, textiles, or plastic lumber, which are durable but also stay in the environment perpetually.[171]

In addition to the small amount of plastic materials that are recycled, an additional 7 percent of plastics are burned as fuels in waste-to-energy combustion (releasing even more greenhouse gases into the environment).

Roughly 30 percent of plastics that have ever been manufactured are still in use (not recycled or discarded), and the remaining 55–60 percent of plastics are thrown away in the trash, where they end up in landfills or the oceans.[169]

The proportion of plastics that are discarded is significantly higher in low- and middle-income countries across South Asia and sub-Saharan African, where recycling infrastructure is low and as much as 90 percent of plastic waste is not properly disposed of and ends up polluting rivers and oceans.[169]

But Why Is This Really So Bad?

Since pictures often speak louder than words can, by using compelling imagery, the documentary *Plastic Paradise* shows how the millions of tons of plastic debris that have entered the oceans make their way to some of the most remote and uninhabited (by humans) islands in the world. One of these islands is the Midway Atoll, a 2.4 square mile (6.2 square km) island in the middle of the Pacific Ocean where roughly 2 million Laysan albatrosses nest.[182] Midway Atoll is home to roughly 70 percent of all the world's Laysan albatrosses.

Unfortunately, this island is now also home to millions of tons of plastic debris.

Every year, an additional 20 million tons of plastic waste end up on the beaches and in the coral reefs and seas surrounding Midway Atoll. As much as 5 million tons of this plastic are fed, unwittingly, to albatross chicks by their parents, who mistake these pieces of plastic for food.[183] Sadly, many of these chicks die from starvation, with stomachs full of plastic rather than food. The pieces of plastic found in their stomachs range in size from as small as a grain of sand (microplastic) to as large as a cigarette lighter.

The Hawaiian Islands, which are a stone's throw from Midway Atoll, are similarly strewn with thousands of tons of plastic trash, which mar the land and seascape. The variety of plastics that wash up on shore is quite astounding, from bottle caps and toys, to fishing gear, computer monitors, and truck tires. As much as 80 percent of "ocean plastics" come from land-based sources. The other 20 percent come from marine sources (including fishing nets, lines, abandoned fishing vessels, or abandoned, lost, or discarded fishing gear).[184] Some estimates place the amount of plastic particles floating in the oceans in the vicinity of 165 million tons (150 million metric tonnes), or roughly 5 trillion individual pieces.[185]

Much of the trash that makes its way to these islands floats along ocean currents called gyres, which move water and debris from big population centers on either side of the Pacific toward the middle of the ocean. When trash passes by the islands, it gets caught, much like with the fingers of a comb.[183]

Every year, plastic waste causes serious damage to marine ecosystems and entangles animals such as dolphins, whales, turtles, and seabirds. The damage caused by plastics may cost as much as $13 billion every year,[186] and this is a conservative estimate. Much of this financial impact is felt in the fishing and tourism industries either by the fouling of fishing equipment, reefs, and other critical habitats, or due to the ugly sight of polluted beaches.[186]

Even worse, as plastics float around the ocean they are exposed to sunlight and ultraviolet (UV) radiation, leading to photodegradation. Photodegradation occurs when UV radiation provides a sufficient amount of energy to allow oxygen into the structures of the plastic polymers. This chemical change causes the plastics to become brittle and break down into smaller and smaller pieces.[187] As the plastic pieces become smaller, they leach chemical by-products – including bisphenol A (BPA), styrene, and polychlorinated biphenyls (PCBs) – into the ocean.[188] Plastics begin to break down within a year of entering the ocean, but the longer plastics remain degrading in the ocean, the more chemicals they leach into the oceans.

And while there are companies working toward straining plastic pieces out of the ocean, too many plastic pieces are too small to be strained out, while others have already sunk to the bottom of the ocean or have traveled to such remote places that trying to achieve this objective is like trying to bail out a sinking ship with a tablespoon.

Even worse, more plastics enter the oceans every year, wreaking havoc in many ways.

Plastic's Effects on Marine Life

Historically, fishing nets were made from grasses, flax, and other plant materials that could naturally biodegrade in the marine environment. Later on, cotton was used, and sometimes wool or silk (which are still natural, biodegradable materials).[189]

Today, fishing nets are made mostly from nylon, a "fiber" made out of plastic materials (for more on this, see Recipe 20 in Part 2). Each year, hundreds of thousands of marine animals become entangled and die as a result of "ghost fishing" by abandoned or lost fishing nets. "Ghost fishing" is the unintentional capture of animals that results from these discarded nets. When fishing gear is left behind, it does not degrade in the same way (or at the same speed) as natural materials. Instead, it persists in the ocean, invisible to animals, and continues to autonomously "capture" fish, sharks, turtles, dolphins, and even whales, as well as other species, drowning and killing them by the hundreds of thousands every year.

Estimates place the amount of abandoned, lost, or otherwise discarded fishing nets and other fishing gear that gets added to the oceans at upward of 640,000 tons each year, and this is increasing.[190,192] "Ghost fishing" continues indefinitely until such fishing gear is physically removed from the ocean.[191]

On top of ghost fishing, discarded fishing gear also alters seabed and marine environments and negatively affects efforts to improve the management of fish stocks.[192]

Plastic is so pervasive and insidious that it has invaded every milliliter (cubic centimeter) and every level of ocean life – from the tiniest plankton to the largest whales.[124] When small fish and micro-organisms consume the smallest pieces of plastics (nanoplastics and microplastics), this can lead to intestinal blockages, illnesses from chemical toxins, or even death, primarily by affecting feeding or foraging habits.[193]

Many fish that have been caught and filleted reveal plastic in their stomachs and intestinal tracts. Seals have been photographed with plastic bags or fishing lines around their necks. Beached whales have been found with over 220 pounds (100 kg) of plastic trash and discarded fishing nets in their stomachs.[194] Hundreds of thousands of marine mammals, fish, seabirds, and turtles die each year from plastic entanglement, ingestion, or starvation.[186,§§§§§§§]

Abandoned or discarded fishing gear takes a toll on and destroys coral reefs by snagging them or breaking them, which can kill them. This is tragic as coral reefs are the second most diverse ecosystems on the planet after tropical rain forests. Thousands of species depend on reefs for their survival. Coral reefs provide protection to coastal areas from storms. They also provide goods and services worth more than $375 billion each year, as well as food for millions of people around the world.[195] Coral reefs also support local economies as a fishing resource and through tourism.

Around the world, reefs have been heavily damaged over the last twenty-five years, and they have been showing further declines in the last ten years. This is in part due to warming waters and in part due

---

§§§§§§§ Starvation occurs as a result of plastic trash remaining in an animal's stomach or intestinal tract and so displacing space for nutritious foods that would normally be consumed.

to ocean acidification (for more details, refer back to Chapter 3). This is also in part due to physical damage from plastics and discarded fishing gear, direct pollution, as well as harmful tourism or overexploitation.[196] Studies show that 75 percent of coral reefs are at risk or threatened from local and global stresses.[197] No one really knows what will happen if we lose all of the coral reefs, but it is probably safe to say that none of us really want to find out.

Thus, there is a twofold problem from plastic debris. The first problem involves the macroplastics, the really big pieces (the buckets, the barrels, the car parts, the nets) that endanger larger marine animals, marine habitats, and corals, as already discussed. The second problem involves the smaller items – plastic bags, broken-down plastics, and microplastics.

Plastic bags, plastic pellets, plastic microbeads (found in certain cosmetic products, but being phased out), and other microplastics make up a large proportion of marine debris (between 77 and 92 percent). Microbeads and microplastics are tiny pieces of plastic, often ranging in size from less than 1 mm in diameter on up to 5 mm in diameter (nanoplastics are defined as plastic particles ranging in size from 1 nm to 1,000 nm). Microplastics and nanoplastics are made from synthetic polymers including polyethylene (PE), polystyrene (PS), or polyethylene terephthalate (PET), and they are small enough to be washed down the drain, where they flow directly into waterways and eventually out into the oceans.[198,199]

Due to their size and similar appearance to "food," microplastics are frequently and unwittingly consumed by many marine animals, including fish, birds, and mammals (including whales). Microplastics (and nanoplastics) are also consumed by some of the smallest marine animals – plankton.[200] These plastics can lead to blocked digestive tracts, malnutrition, starvation, and even death. As many as 80 percent of sea turtles have been found with plastics and other marine debris in their digestive tracts, while roughly one-third of seabird species have also been found with plastic debris in their stomachs.[186,194]

Once in the oceans, microbeads, microplastics, and other plastic pieces may absorb other chemical pollutants that are prevalent in the marine environment, including dichlorodiphenyltrichloroethane (DDT; a previously used insecticide) and PCBs (previously used coolants and lubricants).[201,202] In addition to DDT and PCBs, plastic particles may also transfer other toxic chemicals – persistent organic pollutants (POPs) – into the flesh and fat stores of the animals that consume them.

The level of POPs found in these animals builds up over time in what is known as bioaccumulation.[203] As animals higher up in the food chain – such as carnivorous fish, seals, dolphins, and whales – eat smaller fish (who may have directly consumed these plastics), the toxins and POPs found in the smaller fish become even further concentrated within their own bodies. This phenomenon is known as biomagnification.[203]

Most oceanic predators, and particularly the apex predators (orcas, dolphins, sharks, polar bears, certain seals and sea lions, tuna, salmon, and – finally – humans), have been found to have at least some detectable levels of POPs present in their bodies.[204]

Orcas, specifically those of the Southern Resident subspecies living in the Pacific Northwest, show very high levels of POPs in their bodies.[205] This subspecies is endangered, consisting of only seventy-five members (as of March 2021). In this well-studied population, POP levels in the whales increase with the age of the whales – apart from a few notable exceptions – reflecting the fact that they have had more time to eat and accumulate these toxic chemicals.[206] One exception is that when a whale is pregnant or lactating (producing milk for its calf) or if there is a shortage of prey, circulating levels of POPs increase. In fact, newborn calves and those who are nursing tend to have some of the highest levels of POPs. Why is this? When fat stores break down, they release their accumulated chemical toxins throughout the whale's blood and milk.[205,207] Tragically, these POPs are known to have negative effects on the health (and fertility) of these whales and other marine animals.

POPs may also pose a threat to human health, particularly in areas of the world where fish and other seafood account for 20 percent (or more) of the diet.[111,208] Several studies show that fish species sold in grocery stores and at fish markets around the world contain detectable levels of POPs.[209,*********] These chemicals are not benign – many are known human carcinogens or endocrine disruptors, increasing the risk of cancer, diabetes, obesity, or other chronic illness (see Table 4.1).

********* The POPs most often found in fish are: organochlorine pesticides (OCPs), PCBs, polybrominated diphenyl ethers (PBDEs), perfluorooctanesulfonates (PFOSs), dioxins/furans, and hexabromocyclododecanes (HBCDs).

Table 4.1 Selected persistent organic pollutants (POPs) and their known health effects on the human body

Chemical	Abbreviation	Where it is found/ environmental exposures	Health effects				
			Fetal or reproductive dysfunction	Possible carcinogen	Immune, skin, and/or liver dysfunction	Behavior and nervous system	Endocrine[208]
Bisphenol A[220]	BPA	Plastics and epoxy resins, beverage containers, CDs, toys	X	X	[a]	X	X
Bisphenol S[207]	BPS	Epoxy resin adhesives (close analog to BPA), beverage containers, receipts	X	X		X	X
Dioxins[221]	PCB or PCDD	Insulator fluids, additives to paint and oils. Exposure primarily from high-fat animal products (milk, eggs, meat, fish)	X	X	X	X	X
Polystyrene/ styrene[210]	PS	Widely used in plastics and rubber, packaging materials, drinking cups and other "food use" items	[a]	X	X	X	[a]

Polyethylene terephthalate[222]	PET	Soft drink bottles, water/juice/sports drink bottles, food jars (peanut butter, jelly, pickles, etc.)	X		X	X	X
Phthalates[223]	PHT, PVC, DEHP	Contact with plastic wrap, some children's toys, medical tubing, cosmetics (soaps, shampoos), polyvinyl chloride		X	X	X	X
Polyvinyl chloride[224,225]	PVC	Food packaging, children's toys, shower curtains, building materials (releases dioxins into the environment)		X	X	X	X

[a] Inconclusive results, data insufficient.

What's more, some of these chemicals also affect fertility and fetal development.[210] Therefore, it cannot be overstated that every time we eat fish, shellfish, shrimp, mussels, scallops, or other seafood, we run the very real risk of eating some of these chemicals.

When fish feed on plastic, *we* feed on plastic!

Some 93 percent of Americans, up to 90 percent of Europeans, and 86 percent of teenagers in the UK have detectable levels of BPA in their bodies.[211,212,213] Since World War II, there has been a significant increase in breast cancer incidence as well as a decrease in the age of puberty. While there may be many risk factors attributable to these changes, it is known that both breast cancer and puberty are affected by circulating estrogen levels. It is also known that BPA mimics estrogen, is a carcinogen, and affects fertility, and that the use of plastics has risen 220-fold during this same time period.[214]

BPA is also associated with a greater risk of obesity, heart disease, diabetes, and insulin resistance, and it has been shown to have negative effects on babies, including hyperactivity, anxiety, and depression.[215] BPA may also be linked to polycystic ovary syndrome, premature delivery, asthma, liver dysfunction, immune dysfunction, thyroid dysfunction, and reactivity changes in brain cells.[216] This should be worrying to us all!

We currently are part of a big experiment when it comes to plastic, one that none of us gave our consent to. With scientific clinical studies or experiments, the most basic requirement is to obtain informed consent from subjects. When did you or I give that consent? And now we and marine life are suffering the consequences.

Why Is More Not Being Done to Stop the Use of Plastics and POPs?

One reason for why more is not being done to stop or prevent plastic use is that plastic is a derivative of crude oil. Three of the ten most powerful and profitable corporations in the world are oil and gas companies.[217] Their goal is to make money, not to protect the environment or our health, as evidenced by the oil spills, fracking, methane leaks, and other environmental disasters that occur.

The plastics industry is the third largest manufacturing industry in the United States after steel and cars, and it is responsible for almost

$465 billion in annual shipments. These are powerful interests that enjoy a 2.4 percent growth rate every year.[218]

While it is possible for governments to legislate against the plastic industry, they seem reluctant to do so. In 2020 alone, the United States produced 110–115 billion pounds of new plastic, with an expected growth rate of 2–3 percent each year moving forward. Most of this plastic is used to make disposable, single-use products.[219]

There needs to be an incentive for the industry to change and to find a solution to the plastic and chemical by-product problem. Unfortunately, this incentive does not seem to exist yet, nor does the motivation for the industry to change its ways. The desire for plastic only seems to be growing with time.

While stronger governance may be required to decrease plastic production on a mass scale, there are some things that we as consumers can do to pressure, demand, and affect some of these changes.

For suggestions or "ideas" on what you can do to help stop the flow of plastic into the environment and oceans, please see Recipe 10 in Part 2 of this book.

For more information on the detriments of the POPs previously described in this chapter, please see Table 4.1.

5 ENVIRONMENTAL EXPLOITATION

Our task must be to free ourselves ... by widening our circle of compassion to embrace all living creatures and the whole of nature and its beauty.
—Albert Einstein

Humans are a driving force for life or death, hope or despair, action or inaction.

We can choose to protect species or hasten their extinctions. We can choose to protect the environment or destroy it and the life it supports, including our own. We can choose to slow down the climate crisis and work toward restoring an appropriate balance or we can choose to ignore the damage in front of our eyes and hit a point at which Earth may not remain a healthy planet. We can choose to support organizations that actively protect Earth and its species or we can choose to support organizations that sponsor and encourage the increased use of fossil fuels and plastics or an increase in the hunting, trade, and exploitation of endangered species.

These are the choices each of us makes every day.

If we continue on the paths toward destruction, we may find ourselves on an unrecognizable Earth, where the animals and plants we are accustomed to seeing or coexisting with are vastly diminished or have disappeared entirely.

One of the mechanisms used to protect (or not) wild species is the United Nations Conference of the Parties (CoP) Convention on International Trade in Endangered Species of Wild Fauna and Flora –

also known as CITES.[226] CITES claims to protect over 5,800 animal species and over 30,000 plant species from overexploitation from trade around the world.[227] CITES meets every three years to decide the fate of a number of species. They do this by listing, delisting, upgrading, or downgrading each of these species' levels of protection.

Decisions made at CITES conferences are used to inform the US Endangered Species Act (ESA). The ESA, like CITES, lists, delists, or enacts frameworks designed to conserve and protect endangered and threatened species and their habitats.[228] Species are listed or delisted from the ESA based on how likely they are to become extinct in the foreseeable future.[††††††††]

The stated purpose of CITES and the ESA is "to protect and recover imperiled species and the ecosystems upon which they depend."

While this purpose appears benevolent, in practice it does not always work out that way. Decision-makers are human and often hold a biased and anthropocentric view of the world, essentially saying "humans come first." But who are we to decide the fate of other species? Who are we to decide which species live and which species die? Who are we to decide which species delight us with their spectacle while suffering in silence? Who are we to decide which species thrive (such as domesticated livestock) and which species perish for good (such as the now extinct baiji [Yangtze River] dolphin)?

Sometimes these decisions seem to teeter on an ethical tightrope that too often favors corrupt and wealthy individuals or allows for the organized – yet illegal – trade and black-market sale of endangered species.[229]

As an example, shark fins are a heavily traded "commodity," especially in Asian markets, where they are used to create shark fin soup, a delicacy in China and other parts of Asia. According to many popular media articles, videos, and the like, the fins do not provide flavor to the soup – flavor comes from other ingredients. Rather, shark fins are used *only* for texture, which can now be mimicked by other nonanimal, engineered ingredients (making the use of shark fins an archaic practice). And while the act of shark finning is illegal in many countries, especially those in the Eastern Pacific, shark fins *can* still be

---

†††††††† "Endangered" means a species is in danger of extinction throughout all or a significant portion of its range. "Threatened" means a species is likely to become endangered within the foreseeable future (FWS, 2015).

sold and traded throughout the world, including throughout the United States, essentially giving other countries the green light to continue the practice of shark finning.[230] Tragically, this practice has led to the decline of many oceanic ecosystems as the rapid loss of sharks and other large predators allows invasive species to flourish.[231]

Why should we care?

Well, for purely selfish reasons, we need to keep our oceans healthy, for they provide huge amounts of food for humans. Oceans are also a major source of the oxygen that we breathe.

For moral reasons, many shark species evolved roughly 400 million years ago alongside the dinosaurs, so sharks are a relic of a world we never personally knew. They have survived so much and for so long on Earth – until we came along. Today, many shark species' numbers have declined by 74–92 percent, depending on the species, and many are threatened with extinction.[232] Is this the kind of impact we want to have on the world?

While finning is one of the greatest threats to shark populations, sharks also die in excessive numbers when they get caught in commercial fishing nets as by-catch or when they get tangled up in discarded fishing gear and drown.

Adding insult to injury, many shark species are slow to reproduce, as they are long-lived, late to reproductive age, and frequently reproduce only in smaller numbers. Therefore, allowing any trade in shark fins – whether legal or not – causes irreparable harm not only to sharks, but also to the many other species that depend on the same ecosystems as them.

As an alternative example, when CITES completely and totally banned the trade of all ivory in 1990 – whether legally acquired or not – elephant poaching stopped almost entirely. The complete ban of trade caused ivory prices to plummet and ivory markets to close around the world. These bans meant there was no longer an incentive to poach or hunt elephants.[233] Unfortunately, however, when CITES reversed course in 1997 and allowed some legalized trade of previously acquired elephant tusks, there was a resurgence – actually an increase – in the amount of poaching of this already highly poached species, and this still continues today, leaving both the Savanna African elephant and the forest African elephant (two distinct subspecies) endangered and critically endangered, respectively.[234]

One reason cited for the increase in poaching is that once an elephant is dead and its ivory has changed hands multiple times, it is too difficult to distinguish between previously (and legally) acquired ivory from the past and ivory that has been newly poached and/or illegally obtained.

What these examples demonstrate is that when trade is completely banned and markets are completely closed – when purchasing stops – the incentive to acquire, poach, or kill a species disappears.[235] "When the buying stops, the killing does too."[236]

In the last 100 years, poaching has taken the lives of over 90 percent of African elephants.[‡‡‡‡‡‡‡‡‡]

In 2016, the largest ever continent-wide wildlife survey – The Great Elephant Census – was conducted. The survey found a worse-than-expected decline in African elephants.[237] In less than one decade, 144,000 African elephants (roughly 30 percent of the then-population of African elephants) were killed for their ivory – for their tusks – which are nothing more than overly enlarged incisor teeth.[238] In an odd twist of fate, through "forced evolution," a significantly higher proportion of today's remaining elephants and their offspring are tuskless, which improves their overall chances of survival against the threat of poaching.[239,240]

While poaching is the primary threat elephants face, it is not the *only* threat. Another threat to their survival is human–elephant conflict. This is becoming an increasingly common phenomenon, and it occurs when elephant habitat is destroyed or encroached upon due to the growth in human populations who are competing for the same productive land and water resources.[241]

To address and reduce some of this human–elephant conflict, some regions are adopting strategies such as elephant corridors or the use of beehive-lined fencing, with relative success. Ecological corridors have been shown to facilitate connectivity between herds and enhance gene flow. They also give elephants the space to travel for seasonal migration or for food and water resources. These corridors have been successful in addressing the needs of both humans and elephants.[242] Meanwhile, beehive-lined fences have been shown to deter elephants from entering a family's homestead and eating their crops, improving

---

‡‡‡‡‡‡‡‡‡ One-hundred years ago, there were roughly 5 million African elephants. As of 2020, there are roughly 415,000 African elephants. That's a 91.7 percent loss.

human–elephant relations. Beehive-lined fences also benefit people by providing them with honey and helping pollinate their crops. Both of these mitigation strategies demonstrate how to successfully benefit both humans and elephants at the same time.[243]

Just like humans, elephants are highly social and live in family groups. Elephant herds are led by an elder matriarch (typically the oldest female in the herd) who guides the herd, leads them to food and water, and ensures their safety. Male elephants tend to live with their herds until their young teenage years. At that time, they wander off to be a lone elephant, or occasionally they may form a tight-knit pairing or small group with other males from whom they may learn acceptable social behaviors.[244]

Unfortunately, due to their age, many of these matriarchs and older bull elephants also have very large tusks, making them prime targets of poachers.

Many people who study elephant behavior have seen that after a matriarch dies (or is killed for her tusks) the family herd will often grieve, may fall into disarray, may separate into smaller familial groups, and may have a significantly lower chance of survival than if their matriarch were still alive.[245,246]

Matriarchs spend decades learning from their own mothers the knowledge and experiences that are necessary to locate food and water or to identify predators or safe pathways. Younger family members simply have not had the time to acquire these skills, which puts them and their families at risk. There are documented accounts of elephant calves who have become emotionally traumatized after witnessing the horrific deaths of their mothers. No child – whether human or nonhuman – should ever have to experience that.[247] The hardships an elephant family will go through are only just beginning when their matriarch is brutally taken away from them.

Fortunately, there are organizations in Africa that have done much to protect and defend some of these distressed orphans and elephant families. One such organization is the Sheldrick Wildlife Trust in Kenya, a nonprofit organization that has saved hundreds of elephant orphans who have witnessed horrible tragedy in their very young lives and who would not survive without their helping hands. This organization takes in orphaned elephant calves, rehabilitates them, and eventually returns them to the wild, where several have given birth to their own babies, frequently bringing them back to meet their previous caregivers – a testament to what this organization has done and still does for these beautiful animals.[§§§§§§§§§]

There are many books, documentaries, and films that describe or demonstrate much of these phenomena (see Recipe 21 in Part 2 of this book).[248]

Another large mammal that is struggling even more than elephants is the rhinoceros. Poaching – for their horns – is their primary threat. Traditional Chinese medicine holds the belief that rhinoceros horn has medicinal properties.[249] Yet there is no difference between the keratin in rhinoceros horn and the keratin found in our fingernails. Because of this unfounded belief – one that has no scientific basis – rhinoceros are being killed at alarming rates, and this should cause concern among us all.

At the start of the twentieth century, there were an estimated 500,000 rhinoceros in Asia, Africa, and Europe (but primarily in Africa). Today, it is estimated that there are fewer than 28,000 rhinoceros left, primarily due to poaching.[250] These mammals are long-lived and slow to reproduce. They typically give birth to only one calf every two and a half to five years.[251] Rhinoceros have rather poor vision, making it difficult for them to see (and flee) their foe. While they have been poached to near extinction, the number of rhinoceros that are killed each year is actually increasing.

Clearly this cannot continue.

As recently as 2018, the last male northern white rhinoceros[**********] – called Sudan – passed away from old age. He was under the care and protection of the Ol Pejeta Conservancy in Kenya, twenty-four hours a day, seven days a week.[252] Now that Sudan has died, the northern white rhinoceros subspecies is functionally extinct. There are only two female northern white rhinoceros remaining in the world, and both of them are infertile.

The pangolin, an ant-eating mammal that curls itself into a tight ball when threatened, is the world's most illegally trafficked nonhuman mammal. The pangolin, like the rhinoceros, is being poached to near extinction – in this case for its scales. Pangolin scales, like rhinoceros

---

[§§§§§§§§§] Videos of the Sheldrick Wildlife Trust's orphans and visits from old nursery elephants are available on their website (www.sheldrickwildlifetrust.org).

[**********] The northern white rhinoceros is a subspecies of the white rhinoceros endemic to Africa. Now, only their close cousins – the southern white rhinoceros – remain viable, with approximately 15,000–20,000 individuals remaining.

horn, are made of keratin; and like rhinoceros horn, pangolin scales are used for their supposed medicinal properties in China, Vietnam, and other parts of Southeast Asia.[253]

The giraffe, like the rhinoceros, the pangolin, and the elephant, is also becoming threatened with extinction. In just the past fifteen years, the number of giraffes has declined by more than half. They are being poached for their skin and tails, and they are being killed due to human encroachment on (and conflict over) their habitat.[254] Over the last decade, more than 40,000 giraffe skins and body parts have been imported into the United States from Africa to make taxidermy juvenile giraffes, custom giraffe jackets, rugs, giraffe-leather bible covers, and bone carvings.[255] The desire for giraffe parts resulted in the International Union for Conservation of Nature (IUCN) placing giraffes on the international endangered species list for the first time in 2016, changing their status from "least concern" to "vulnerable" and listing two subspecies as "endangered."[255,256]

Tigers represent yet another critically endangered species that is rapidly being killed off for use in traditional Chinese medicine. Tiger bones, despite a lack of any evidence for their effectiveness, are being used to treat ulcers, typhoid, malaria, dysentery, burns, and even rheumatism.[257] Other tiger parts, such as their whiskers, are worn as talismans or as protective charms, while tiger skins are valued as trophies or are worn as a symbol of wealth.[257]

Of the original nine subspecies of tigers, three have become extinct in the last eighty years. It is predicted that tigers could become extinct in the wild within the next decade if nothing changes.[258] In the early 1900s, there were over 100,000 tigers living in the wild. Today that number has plummeted to fewer than 3,900. There are actually *more* tigers in captivity today (roughly 5,000) than there are living in the wild. Tigers in captivity are primarily raised to supply bones for traditional medicine or for the illegal wildlife trade.[259,260] In fact, only an estimated 6 percent of tigers in captivity live in accredited zoos or similar facilities. The vast majority "live" in private homes, private "zoos," or sideshow petting parks or are used in traveling or roadshow circuses – which are notorious for abusing their animals.[260,261,262]

In Vietnam and China, more than 1,000 bears – mostly Asiatic black bears and sun bears – are confined to small cages where catheters are placed into their gall bladders to extract bile. Bear bile has been used in traditional Asian medicine to treat liver and kidney disease. Bear bile is now even being used as a "possible treatment" for COVID-19 (despite the fact that there is absolutely no proven efficacy).[263,264]

The bears used in these "bile farms" often spend their entire lives confined to a tiny cage, which may not even be big enough for them to turn around in. Many of these bears are deprived of any kind of mental or physical stimulation and often succumb to infection by the time they are only four to five years old. For reference, the average lifespan of these bear species in the wild is roughly twenty-five years.[265]

While the demand for bear bile has declined in the last few years, most bears used in these operations are not released back into the wild, and instead are starved to death or are killed to sell their body parts. And because breeding bears in captivity has proven difficult and often unsuccessful, many of the bile farms that do still exist will often restock their farms with bears from the wild.[265]

Even endangered species such as sea turtles are not safe from the illegal wildlife trade. All species of sea turtle are listed as endangered. Yet China – which is responsible for 98 percent of all trade in sea turtles – still produces and sells products made from sea turtles (which are highly prized as ornaments or as ingredients in traditional medicine).[266,267]

There are hundreds of other examples of animals being killed, maimed, poached, or even tortured so that humans can consume them, whether as medicine, food, art, or live entertainment. Traveling circuses have a history of using wild animals as spectacles, but there is also a history of wild animals being used as entertainment in aquariums, dolphinariums, or other shows around the world that appear benign.

Each year, millions of people flock to places that house captive dolphins, orcas (killer whales), and other sea animals. What all of these places have in common is that they use these animals for entertainment, often in shows that display the animals performing unnatural behaviors, including playing with balls, jumping through hoops, or carrying a person on their bodies.†††††††††† During the day, these animals perform a handful of shows and may even appear "happy." But after everyone has left for the day, sight unseen, they are often kept in small enclosures with barely enough room to swim, in unnatural environments, and often with other animals that are not even part of the same family group

---

†††††††††† For some examples, I suggest searching online for "traveling dolphin show" and watching some of the video results.

or even, at times, the same species. These conditions are enough to drive some animals mad.[268,269,‡‡‡‡‡‡‡‡‡‡]

Likewise, millions of people visit zoos around the world every year to observe elephants, lions, tigers, pandas, polar bears, and other wild animals. Frequently, these animals are seen sleeping, pacing, or bobbing back and forth in what are known as stereotypic behaviors, or signs of stress.[270] Even in the best zoos, where animals have more space to roam, these behaviors have been seen. This may not come as too much of a surprise when we think about some of our own behaviors during the lockdowns of the COVID-19 pandemic, where many of us were forced into seeing the same surroundings day after day after day. The lack of social stimulation and exercise during the lockdowns led to declines in mental health and increased levels of anxiety and stress, so it is not difficult to imagine that this very well may be what animals in captivity experience *all* of the time.

While many zoos and aquariums posit that they help educate and conserve animal species – and for some species they possibly do (e.g. those that would be extinct in the wild) – it is difficult to see how they are doing enough to protect species when so many of the animals in their care die younger than their counterparts do in the wild, often as a result of changes to their immune systems, behavior, or physiology.[271,272] Moreover, try as they might to increase their numbers through captive breeding only, some zoos and aquariums – especially those that are less reputable – still source some animals from the wild.[273,274] In Chapter 6, I address in more depth some of the ways in which wild animals are captured, as well as the captive entertainment industry.

I would like you to pause for just a moment and consider why so many animals are being treated in these ways.

Some reasons may be related to ignorance or a lack of awareness, such as human population growth unintentionally pushing nonhuman animals out of shared land. Other reasons may be far more insidious, such as purposefully seeking out animals to be used for medicines or entertainment. And finally, other reasons may simply be related to corruption, as in the following two examples.

---

‡‡‡‡‡‡‡‡‡‡ I recommend watching the film *Blackfish* for a better understanding and visual depiction of this description.

The first example is in the Sea of Cortez, a very small 160,000 km² waterway separating the Baja California peninsula from the Mexican mainland. The Sea of Cortez is home to a small porpoise species called the vaquita. The Sea of Cortez is the vaquita's only habitat, and it is the only place they have ever lived.

Tragically, the vaquita is nearly extinct, with fewer than twenty individuals remaining.[275] The cause of their near-extinction is an incidental by-product of greed for another endangered species that shares the same habitat, the totoaba. The totoaba is a large species of fish that is prized in China for its swim bladder, which is (again) believed to have medicinal powers.[276] Totoaba are caught – illegally – with gill nets placed throughout the Sea of Cortez by totoaba poachers. Gill nets are essentially a curtain of netting designed to capture and kill everything that swims into them, including the totoaba. But since gill nets are indiscriminate, they also capture and kill animals such as sea turtles, sharks, and the vaquita.[277]

In this example, two critically endangered species – the totoaba and the vaquita – are at risk of extinction; one for its perceived medicinal qualities and one by accident. There are groups – including Sea Shepherd – that are trying to protect and save vaquitas and their habitat. Sadly, these organizations are greatly outnumbered by poachers and others who continue to capture and trade the totoaba.

The next example is particularly tragic in that it is a response to the desperation some are feeling as a result of the COVID-19 pandemic.

While there have been *some* positive responses and outcomes to the pandemic, including clearer, cleaner skies in large cities in rich countries, giving some wildlife space they can use to thrive, in some poorer countries, some people are being driven to extremes to support themselves and their families by poaching.[20] This is happening because all of a sudden millions of people are unemployed and have nothing to fall back on. There are very few, if any, social safety nets or coping strategies available in these areas. This has also led to more urban-to-rural migrations, where daily wage earners feel that they have no other option besides poaching, logging, or other harmful activities to feed their families.[20]

Bushmeat,§§§§§§§§§§ rhinoceros, and elephant ivory poaching has increased in Kenya and Cambodia during the COVID-19 pandemic,

§§§§§§§§§§ Bushmeat is a catchall phrase for the meat of wild animals, but most often refers to the remains of animals killed in the forests and savannas of Africa, including bats, monkeys, rats, snakes, and other wild animals.

as has tropical deforestation, agricultural expansion, and illegal mining in Brazil, Colombia, and Cambodia.[21]

Restrictions on international travel have effectively shut down tourism, leaving many wildlife parks closed, diverting law enforcement away from ranger patrolling and instead toward COVID-19-related activities, increasing the risk and exposure of wild animals to poaching.[20] Furthermore, when people increase their contact with wild animals, they increase their risk for future pandemics.[21]

Humans have a great capacity for love, empathy, and stewardship, emotions that can also be seen in other animals (just look at your cat or dog!). Yet the previous examples indicate that when we act out of desperation or self-interest, humans also have a great capacity to harm or destroy.

To avoid additional ongoing destruction of wildlife in the immediate term, interventions such as cash transfers or the distribution of food parcels are likely to be needed in rural areas. In the longer term, income diversification, safety nets, and other coping strategies are needed so that communities are not solely dependent on – as just one example – ecotourism for their livelihoods.[20]

Currently, rates of extinction for various species are anywhere from 10 to 1,000 times higher than would be expected without human influence.[278] The natural or background rate of extinction is measured for a specific class of animal over a specific period of time; for example, "one species of bird (classification) is expected to go extinct every 400 years by natural causes."[279] Mammals, on the other hand, have an average species lifespan of 1 million years, and their natural background rate of extinction is one species of mammal (classification) every 200 years (due to natural causes).[**********] Yet in just the last 400 years more than 89 mammal species have gone extinct – an extinction rate that is at least 45 times faster than expected.[279,280,281]

Today, we are seeing that our direct and indirect actions against animals, their habitats, and the environment contribute to – and may be responsible for – these precipitous declines in the numbers of both individuals within a species and entire species themselves.[282] Each year, more species – whether plants, birds, mammals, reptiles, or amphibians – become threatened or vulnerable to extinction.

[**********] www.pbs.org/wgbh/evolution/library/03/2/l_032_04.html

Each species that goes extinct had a role in its ecosystem. Each species that goes extinct was composed of individuals who each had their own life story. Each species that goes extinct represents millions – if not billions – of years of unique evolutionary history and phylogenetic diversity.[280,283,]††††††††††† Let us not forget what we lose when a species goes extinct.

Tragically, a very high proportion of animal extinctions have been in mammals, representing a tremendous loss of phylogenetic diversity (mammals represent a disproportionately high share of phylogenetic diversity). Mammals also represent a large proportion of the currently threatened species.[283] However, this should not come as a surprise when looking back at history. In the past, human-caused extinctions have been size-based, devastating the largest of the mammals first, thereby eliminating the important ecosystem functions and services they provide.[283]

Sadly, the IUCN predicts that we will lose 99.9 percent of critically endangered species and 67.0 percent of endangered species within the next 100 years, many of them mammals.[283] This means that cows could – one day soon – be the largest land species remaining on Earth.[284]

To put this information into perspective, of all life on Earth (measured as biomass), animals make up only a tiny fraction: 0.4 percent of the total biomass.‡‡‡‡‡‡‡‡‡‡‡ Mammals make up only 7.0 percent of this 0.4 percent animal biomass – or put another way, mammals make up just 0.028 percent of all the biomass that is on Earth. Of this 0.028 percent of mammal biomass, wild mammals make up less than 4 percent – or 0.001 percent of Earth's total biomass. This means for every 100,000 pounds of biomass on Earth, a mammal represents only 1 pound (see Figure 5.1).

However, this number wasn't always *this* low. Figure 5.1 shows how wild-mammal biomass levels (as compared with human

---

†††††††††† Phylogenetic diversity is the amount of independent evolution among various species – in other words, phylogenetic diversity represents the characteristics (either physical or genetic) that allow for the groupings of and distinctions among various species. The amount of distance between various branches of a phylogenetic tree represents the amount of time passed in the evolutionary development of a new species.

‡‡‡‡‡‡‡‡‡‡ Biomass (also known as biological mass) is the total mass of living organisms in a given area or volume.

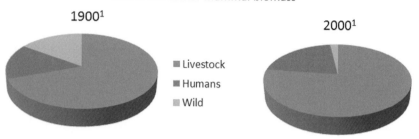

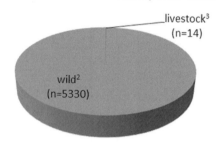

Figure 5.1 Comparison of wild mammal biomass versus human and livestock biomass

Image credit: Zeller, Starik, & Gottert (2017)

and livestock biomass levels) have shifted between the years 1900 and 2000.[285,286]

The third pie chart in Figure 5.1 represents where genetic diversity resides among all mammal species. This bottom image makes very clear that nearly all genetic diversity exists in wild mammal species (which make up less than 4 percent of mammal biomass) as compared with the less than 1 percent of genetic diversity among livestock (which make up more than 75 percent of all mammal biomass).[287]

Additional factors that speed up extinction rates are our desire for animal products, including meat, dairy, eggs, leather, skins, bones, furs, ivory, scales, and other animal parts, which are immensely damaging to species, their habitats, and the environment.

Our taste for meat and dairy products destroys pristine habitats all over the world, and especially in forested areas, to create more land for domesticated animal grazing or to grow feed.[288] The loss of forests and wetlands affects microclimate weather patterns, changes the overall environment, and hastens climate change by removing or reducing carbon sinks and allowing more carbon dioxide to accumulate in the atmosphere.[289] Warming temperatures brought on by greenhouse gas emissions and climate change pose additional challenges to species that already live at the extremes or edges of their temperature or habitat ranges, putting them and possibly ourselves at greater risk of hunger, disease, death, or extinction.[290]

Our desire for animal parts, whether for decoration, food, or as perceived natural remedies (most of which are unproven), puts us into close contact with many species. In addition to the effects that our actions have on the environment, our interactions with other species increase the risk that we will come into contact with new infectious diseases, including viruses that we do not have immunity to, some of which may kill hundreds of thousands if not millions of people (e.g. COVID-19).[291] We are seeing some of the harmful effects that our exploitative interactions with wild animals have both on them and on human life. And despite the numbers of people falling ill and dying from COVID-19, only time will tell whether or not we have learned our lesson.[292]

Maybe we haven't: After only a few months of closure, the wet market in Wuhan where COVID-19 is thought to have originated reopened, and this is not the first time that this has happened.[293] Studies and history show that in past disease outbreaks wet markets were shut down in the weeks that followed, only to be reopened a few weeks or months later.[294,295] This irresponsible cycle continues.

It seems time to finally learn the lesson that high-risk, high-contact markets or other sites with a high propensity for wildlife–human contact need to be shut down permanently, thereby removing at least one major risk factor for the initiation and spread of future novel infectious diseases.[6] It also seems time that we work with individuals and families to identify upstream approaches and adaptive strategies that will help prevent them from feeling the need to resort to destructive activities for their own survival – whether wildlife poaching or deforestation.

We can choose to be on the right side of history and ban the trade of endangered animals and animals that may become endangered in the future. We can also choose to end our exploitative and harmful treatment of animals. I believe these are necessary steps that will decrease the threat and prevent the spread of new infectious diseases and at the same time increase the resilience and strength of the planet's ecosystems. I fear that if we do not make these changes we could eventually pay the ultimate price.

6 SPECIES EXPLOITATION FOR ENTERTAINMENT

The greatness of a nation and its moral progress can be judged by the
way its animals are treated.
　　　—Mahatma Gandhi

While this chapter may feel a bit tangential, in reality, it is as
relevant as all of the other areas addressed in this book. As a species,
humans have polluted the Earth with our fossil fuels, our plastic, and our
dietary habits. We have destroyed many habitats, wiped out large
numbers of wildlife, and made it increasingly difficult for many species
to survive, let alone thrive. To top it all off, we have also taken in members
of some species – including those that are critically endangered – and
forced them to live in horrific conditions, where they are physically and/or
psychologically abused or starved as a way to "motivate" them to work.

These are the conditions many individuals live in at some
animal amusement parks, circuses, sideshows and roadshows and at
certain tourist attractions.

To better help you understand some of these settings, I give a
few examples in the following pages. Some of the examples describe
situations where food is withheld from animals until they are performing in a show. During shows they will be fed "rewards" – often called
"positive reinforcement" by their trainers.§§§§§§§§§§§

Other examples include animals who are chained so tightly
during the day or night – on the legs, through their snouts, or

---

§§§§§§§§§§§ Much of this has been outlined in books and documentaries about orca and
dolphin captivity, elephant tourism in Southeast Asia, and circuses.

elsewhere – that they bear horrific scars wherever they are tethered. ************ Some examples are of animals who are force-fed, or who are injected with sedatives so that tourists can take photographs with them. These animals also have their claws or teeth removed so that they cannot fight their oppressors.††††††††††††

There are many examples of animals being deprived of all basic physiological and psychological needs. They are deprived of food, shelter, warmth, kindness, and all normal behaviors. Do enough research and watch enough videos and you will begin asking yourself: Is this the legacy we want to pass on to our children and grandchildren?

I believe that in addition to protecting ourselves, it is necessary to help and protect the other species that live on this planet with us, and we need to offer them the same protections that we want for ourselves. We do not live in a vacuum, and it is important to "speak out [and do] for those who cannot speak [or do] for themselves."‡‡‡‡‡‡‡‡‡‡‡‡

As I have said before, being resilient to climate change, food insecurity, and environmental uncertainty depends on the health of this planet and *all* of the species on this planet. Having too few healthy species in too few healthy habitats will increase the risks to us all.

This chapter focuses on captive animal welfare. Based on a few specific supporting examples, I believe that the majority of zoos, aquariums, and amusement parks that house captive animals to entertain humans do not actively contribute to conservation, though most of them will claim that they do.

Example 1 Tilikum

I strongly encourage you to watch the documentary *Blackfish*. The film goes into significantly more detail about these circumstances than I will here. But in short, SeaWorld (and other similar marine amusement parks) house a number of marine animals taken from the wild. These animals are forced to live very unnatural lives in very unnatural conditions.

The film's focus is on the orca Tilikum, the largest male orca ever held captive at SeaWorld Orlando (Florida). He was stolen from the wild –

*********** This can be seen clearly in videos on elephant captivity or "performing" bears.

†††††††††††† This is best demonstrated by videos of tigers in captivity who have had their claws and teeth removed and are sedated to allow tourists to take photographs with them. These animals are also frequently whipped or harassed.

‡‡‡‡‡‡‡‡‡‡‡‡ This is a beautiful quote from Dr. Jane Goodall.

from the waters of Iceland – in 1983 at the age of two.[248,269] At that age, he would have been completely dependent on his mother for milk, food, socialization, and basic survival. After his capture, he was held for a year in a tiny cement holding tank in Iceland, where he learned how to survive on dead fish and where all he could do all day was float on the water's surface and perhaps swim in small circles.

A year later, he was flown to a marine amusement park off the western coast of Canada, near Vancouver Island. Instead of being surrounded by his mother and other family members in the open ocean, he was held captive for fourteen hours a day in a tiny tin can of a cell, with nowhere to go, no access to natural light, and nothing to do but be attacked, tortured, and bullied by the two female orcas he was held captive with.[248,269]

During the day, Tilikum would be released from that tiny cell into a slightly larger pool to perform for paying customers. During training sessions, food was withheld. Aggression among the three orcas was common. Finally, in 1991, after seven years of this treatment, day in and day out, Tilikum (and his cellmates) killed a young woman after she fell into their pool by holding her underwater.[§§§§§§§§§§]

In 1992, after this incident, Tilikum was sold to SeaWorld Orlando, where he became nothing more than a sperm bank who was also used to perform one specific trick at the end of each show. Once again, Tilikum was often confined to a small medicine pool hidden behind the larger performance pool, where all he could do was float at the surface and await his turn to "perform."

Tragically, in 2010, after eighteen years at SeaWorld Orlando, Tilikum killed again. This is the story that is well documented in the film *Blackfish*. The film also presents many of his former trainers, who describe training methods that included withholding food or keeping Tilikum enclosed in small pools or cells where other whales could attack him.[*************]
Tilikum held on to "life" as long as he could, but eventually he died on Friday, January 6, 2017. He became a nonhuman spokesperson for the

---

§§§§§§§§§§ What makes this so unique and unusual is that there has never been a documented case of an orca killing a human in the wild.

************* Many orca societies are matrilineal (female-based, like elephants are). Females are the boss. When a female attacks a male, despite his size, there is not much he can do. Additionally, because Tilikum was taken at the age of two, his emotional development was stunted, which may have allowed these attacks to continue due to a victim mentality. There is much more information on this available at https://theorcaproject.wordpress.com/links, and please also see the research done by Dr. Lori Marino on whale cognition at https://journals.plos.org/plosbiology/article?id=10.1371/journal.pbio.0050139.

plight and cruelty faced by dolphins, orcas, and any sentient animal in captivity.

Tilikum's story demonstrates that when you remove an animal from its natural environment, where he might normally swim 75 miles (120 km) each day in search of food or mates, to a wholly unnatural environment – a concrete swimming pool – it eventually becomes too debilitating, both physically and psychologically, to keep on living.[269]

There are those who will argue that Tilikum received the best veterinary care available, but in the wild, orcas often live longer and healthier in their own family groups. No veterinary care is required.

There were many opportunities, organizations, and donors willing to assist Tilikum to become a free whale once again – or at least a retired whale who could have lived out his days in a sea pen, a setting where he could have felt the natural rhythms of the ocean, been more in control, and made some of his own decisions once again. Over and over he was denied this right. Instead, he languished in pools, some barely as large as a backyard pool, where he logged (floated) at the surface for too many years.

During the lockdowns of the COVID-19 pandemic, many of us circulated memes that spoke of the tedious boredom of being stuck in our own homes for weeks at a time. Many of us had little opportunity for movement or activity, and we quickly grew tired of the same television shows. These experiences have been terrible for our mental health. But imagine being Tilikum or any other captive orca, dolphin, cheetah, elephant, or bear and being stuck in the same tank, small zoo "habitat," or cage for years at a time, and maybe – just maybe – you will understand the insanity that many of these creatures exhibit.[296,297]

Today, there are still at least sixty orcas held in captivity around the world. Some countries, including China, are building new aquariums so that they can house additional orcas for their entertainment industry.[298]

There also are nearly 3,000 dolphins held in captivity in aquariums, zoos, and marine parks around the world, with new aquariums similarly being built to hold more dolphins for entertaining humans.[299]

How these innocent mammals – who are so like us in intelligence – are rounded up and stolen from their homes, their families, and their freedom and the conditions in which they are forced to live demonstrate how cruel, inhumane, and all-about-the-money these amusement parks are. However, they also highlight these mammals' drive to live.

Tilikum's real life ended at the age of two when he was stolen from his family. Everything, including his life, was stolen from him. He became quite literally nothing more than a shell of a whale.

When I first learned of Tilikum's story, my son was exactly the same age as Tilikum was when he was stolen. I imagine his story as being my son's story, or the story of other young people. Given the separation he went through, the abuse from other whales and humans, the lack of proper socialization, and the constant isolation he was forced to live in, it is not so hard to imagine how he became the "whale" that he did, a tortured soul who sadly killed and eventually died due to the pain and suffering inflicted on him.

Tilikum's story, among other reasons, is precisely why I support ecotourism rather than tourism that involves the exploitation and/or emotional or physical abuse of animals (I will discuss more about the benefits of ecotourism in Part 2 of this book). This brings me to the next, very similar example of Tokitae (Lolita) at the Miami Seaquarium, whose story is just as heartbreaking as Tilikum's.

Example 2 Tokitae (Lolita)

Tokitae (who has also become known as Lolita) is the only orca at the Miami (Florida) Seaquarium, and she has "lived" there for over fifty years. Tokitae has been the *only* orca at the Seaquarium for over forty years.[300] She inhabits the smallest orca tank in the world.[†††††††††††††]

Her story of captivity begins in August 1970, when she was stolen from Puget Sound, Washington. Like Tilikum, she was very young, being only four years old when she was taken. On September 24, 1970, she was transferred to the Miami Seaquarium, where she shared her tiny tank with another orca – named Hugo – who was a member of the same family group. The reason why so much is known about Tokitae (and Hugo) is that her family – the L-pod of the Southern Resident killer whale population – is extremely well-studied and has been observed by researchers for decades.

In 1980, Hugo died from a brain aneurysm, and since his death, Tokitae has been the *only* orca at the Miami Seaquarium.[301]

[†††††††††††††] To see how small Tokitae's tank is relative to her body, watch this video: https://youtu.be/-v5lH3HeWS0. Tokitae is roughly 21 feet (6.5 meters) long and swims in a tank that is 80 feet (24 meters) long, 35 feet (10.6 meters) wide, and 20 feet (6 meters) deep (at their longest, widest, and deepest points). That would be like you or me swimming in a pool that is 22.2 feet (6.7 meters) long, 9.7 feet (3 meters) wide, and 5.5 feet (1.7 meters) deep.

She currently shares her (already too-small) tank with a pair of Pacific white-sided dolphins, and at age fifty-six she continues to perform the same mundane and unnatural tricks she was taught many years ago. When you consider how intelligent she is, being able to understand the signals that humans give, you gain a deeper understanding of how the tediousness and repetition of her daily life may very well be enough to make her (or anyone) go crazy.

The indigenous Lummi Nation – who live on the shores of Puget Sound – and other organizations have been fighting in court to get Tokitae released from the Miami Seaquarium. Their hope is that she can be rehabilitated, "untrained," and released back to her family, who still swim through the Salish Sea of Puget Sound on an annual basis. Among the many ethical reasons for releasing her from a life of confinement is that Tokitae's mother is still alive, and she still swims as a member of the L-pod. These particular orcas are matriarchal, and they stay with their families for life. It is expected – by many researchers who study this group of whales – that Tokitae would be accepted back into her family pod, where she could live out her remaining days as a wild orca.[302]

Instead, the Miami Seaquarium's owners have refused to release Tokitae to be rehabilitated. She is the major draw to the Seaquarium, and her daily shows bring people in to spend their money. They treat her as a commodity. This is evidenced by the fact that in 2017, when Hurricane Irma passed over Miami, she was left alone in her tiny tank (it is unclear whether her trainers stayed at the Seaquarium to attend to her) without shelter and with nowhere to dive away from the storm.[‡‡‡‡‡‡‡‡‡‡‡‡‡] It is a miracle that she survived.

There are also hundreds of dolphins in captivity and, like orcas, they too suffer in silence.

Example 3 Dolphin Captivity

There are many places around the world where people pay to get into the water with dolphins living in captivity. Often, these dolphins will pull people through the water while they are holding on to their dorsal fins. Other times, people will pose for photographs next to or on top of the dolphins themselves.

While this may feel like a dream to those people who want to have this experience, I am sure it is not a dream for the dolphins who are forced into it.

---

‡‡‡‡‡‡‡‡‡‡‡‡‡ www.miaminewtimes.com/news/orca-lolita-left-outside-during-irma-9665020

In their natural environment, such as the ocean, if a dolphin wants to interact with or learn about people, it will approach them of its own volition. But in a tank, dolphins are forced into interacting with tourists or else they may not be fed. There are reports and video footage of dolphins lashing out at visitors or their trainers in harmful or painful ways.[303],§§§§§§§§§§§§§ Dolphins experience a lot of stress from these interactions and from having their freedoms taken away. There is a significant amount of literature on stress in dolphins and it all shows that captivity is harmful and debilitating and frequently leads to disease and early death.

Of the dolphins that survive the capture process, 53 percent of them die within 90 days. It is also estimated that every seven years half of *all* captive dolphins die as a result of intestinal disease, chlorine poisoning, or stress-related illnesses, including ulcers and/or bacterial infections.[304] The naive tourist will never be made aware of this, as there are many dolphins available to replace those who die.

Speaking of how dolphins are acquired for the captivity industry, I strongly recommend that you watch the film *The Cove*.[305] While there are some very difficult scenes to watch, it exposes the brutality of the dolphin hunts (and captures) that occur every year in Taiji, Japan. This is the place from where many captive dolphin facilities acquire their dolphins.[306]

Each year, these dolphin hunts start on September 1 and go on until March 1, so if you want to see for yourself how these captures are done, I urge you to search for Ric O'Barry's "Dolphin Project" pages on Facebook or Twitter and watch the livestream videos for yourself. What you will see is dolphins being hunted relentlessly, sometimes for hours at a time, so that they can be captured for marine animal amusement parks, or else be mercilessly murdered in order to be eaten.***************

I will admit, *The Cove* and these livestreams ignite a fire in me to talk about the things I have observed and to teach others about the tragedies and horrors of animal exploitation. The following is just one recollection from a livestream I observed a few years back, one that I was fortunate enough to have posted on *Huffington Post* when I was a contributor.†††††††††††††

> Yesterday, a very small pod of Risso's dolphins, maybe five, or seven, but no more than ten, were brutalized in a five-hour

---

§§§§§§§§§§§§§ www.dolphinproject.com/campaigns/captivity-industry/swimming-with-dolphins

*************** It is worth stating that many dolphin species contain high levels of mercury and other toxic compounds in their flesh. When people consume dolphin meat, they also consume these toxins (much as described in the Chapter 4).

††††††††††††† If you want to read more about this, search online for "Dana Hunnes, Huff Post, Families fight for their lives."

drive-hunt towards the cove of Taiji. These dolphins fought for their lives, repeatedly holding their breath while exhausted, swimming under the hunters' boats, and trying to outpace the them.

Eventually, their exhaustion, their confusion, and their disorientation by the loud banging of the boats betrayed them as they began swimming towards the cove instead of away from it towards open water. However, the unrelenting cruelty and greed of the dolphin hunters in Taiji cannot be understated.

The going price for a slaughtered dolphin is perhaps $250–300, in fact, the amount of fuel used to chase this small pod of dolphins into the killing cove likely cost more than the dolphins themselves were worth to the hunters, dead. In watching the video, it almost started to look as though the hunters got a thrill from terrorizing these dolphins, laughing, smoking, and just having a grand old time of it.

So much energy, so much hate, so much time, so much anger directed at such a small, harmless, innocent, beautiful pod of Risso's dolphins, some of the shiest, and most intelligent creatures on the planet whose only wrong move was that they swam by this place at the wrong time.

These hunts take place during very specific times of the year, during the migration of dolphins near the coast of Japan where they are easy to spot.[307] These dolphins are swimming to their next destination in search of food, mates, or socialization.

In this pod was a tiny baby, probably no more than a few weeks old, so dependent on her mother for everything, milk, care, navigation, everything. It was obvious how much this little family wanted to live and keep each other, especially the baby, safe from harm, just like any other family would, human or nonhuman.

They swam under the loud boats, tried to evade the wall of sound that surrounded them, and tried to escape until they no longer could. Each and every time they surfaced to catch a necessary breath, to fill their lungs and their over-exhausted and fatigued muscles with oxygen, the hunters ran their boats at full throttle after them, for more than five hours.

You could hear in the hunters' voices, see in their actions that they were getting angry and impatient. Finally, when the hunters could take it no longer they used nets which they extended from boat to boat under the water to block the dolphins' escape route and herd them into the killing cove.

And in a last act of cruelty, they killed every one of the pod members, every dolphin, except the one small baby, which they manhandled, put into a skiff, covered with a tarp, and drove back out to open ocean, alone, with no mother, with no other pod-member.

Even rewriting about this now brings tears to my eyes, as this baby had no way to defend itself, no way to obtain food, milk, or water. In the best-case scenario, this baby was adopted by another pod, but that is extremely rare. Most likely, this Risso's dolphin baby died of starvation or predation.

The same behaviors occur with pods of larger dolphins (typically bottle-nose dolphins and a few other species), which are often taken to marine animal parks for entertainment, except rather than kill members of these species, those individuals that are not selected for captivity are often driven back out to the ocean having lost family members, pod-mates, and a part of their very essence.

It is for these very reasons that I cannot step foot inside a marine amusement park. It is for these very reasons that I write about this issue, talk about it to others, and try to educate about and create awareness of it.

So much is known about the intelligence and social lives of dolphins that treating them as mere commodities and forcing them to live in small pools and do tricks or starve should be things of the past. Many studies have been conducted on dolphins and their abilities, their brains, and their social lives (see the recommended books and films at the end of Part 2 of this book) – we know that to continue exploiting them is simply immoral.

If you ever catch yourself asking, "How do we know dolphins are smart?" then consider this: dolphins have learned to read humans. Dolphins can understand human signs, symbols, and sounds – this has been documented. Yet we have not even begun to break the code of their language or their understanding of the world.[308] Many dolphin species are under threat by humans, whether from hunting or pollution or as by-catch.[309] It is past time to end dolphin and whale captivity, as well as the captivity of other species, including elephants.

Example 4 Captive Working Elephants

Although there are many other examples I *could* discuss, I will only present one more, and that is of captive working elephants.

Elephants, like humans and dolphins, are very self-aware, intelligent, sentient, and family-oriented, and they do not belong in captivity or as "working" elephants.

Any time a tourist pays for an elephant ride or "experience" – most frequently in developing countries (although this also happens in developed countries, such as the United States) – they are paying for the torture of an innocent life. Wild elephants do not know how to "give a human a ride on their back." This is something they are trained to do using methods that involve horrific abuse and torture.

To begin the training or "taming" process, a baby elephant is stolen from the wild – from its family. Adult elephants are far too large to initiate in the training process, which is why babies are taken. For days on end, baby elephants are tied up, beaten, starved, and tortured until they are broken. This is known as "the crush."[310],‡‡‡‡‡‡‡‡‡‡‡‡‡‡

Once the crush is finally over and the elephant is "tamed," this elephant will often follow the same life trajectory as the other 3,800 captive elephants in Thailand and the thousands more throughout Southeast Asia and the rest of the world. These stolen elephants perform in shows until they are roughly ten years old, and then they become riding elephants. Tourists pay to sit on a bench that is strapped to the backs of these elephants. They are forced to give several rides each day, and when they are not performing or giving rides, they spend most of their lives tethered to a chain inside of a small stall, or else they may be tied to a tree or post with no freedom of any kind.[311]

Other elephants may be working elephants. Rather than being forced to give rides to people, they haul heavy logs through the forest or carry hundreds of pounds of decorations during festivals and celebrations. Many Hindu temples in Kerala, India, own elephants. These elephants are exploited, shackled, tortured, or forced to carry hundreds of pounds of weight, and they often die young from maltreatment and/or stress.[312] A film that provides many details about these phenomena is *Gods in Shackles*.

---

‡‡‡‡‡‡‡‡‡‡‡‡‡‡ During the crush, elephants aged three to six years (and sometimes younger) are repeatedly beaten with sharp tools, constantly yelled at, stabbed, burned, beaten, starved of food, and deprived of water. Bull hooks are used to stab the elephant, tug its ears, and control it. When an elephant finally gives in to the torture and loses all of its former self, having lost the will to fight back, the process is done, but the anguish and fear never go away. For the rest of their lives, these elephants will be under the control of fear, pain, bull hooks, or beatings. For more information on this, search online for "elephant crush" to see this process for yourself.

Throughout the world, there are many examples of the maltreatment of wild animals. It often occurs behind the scenes, when no one is watching, at circuses, zoos, festivals, sideshows, and in private homes. It happens to a range of animals, from dolphins and orcas to elephants, tigers, and bears, and even macaques, orangutans, and gorillas.

This is happening right now, but it shouldn't be. Every animal has the right to its own life and freedom. We can change this, and I discuss where this change is happening in Chapter 7.

THE POSITIVES
7 Examples of the "Good" Being Done around the World

Do not be dismayed by the brokenness in the world. All things break. And all things can be mended. Not with time, as they say, but with intention. So go. The broken world waits in darkness for the light that is you.
　　　　　—L. R. Knost

While the first six chapters of this book have given an overview of how much harm humans are causing to the environment, other species, and ourselves, it is also necessary to acknowledge examples of the good and beneficial acts that many of us already engage in.

I will be the first to admit that it is easy to lament and feel hopeless about all of the environmental degradation and species exploitation that is happening (Chapters 5 and 6), to feel fear over the ramifications of overfishing (Chapter 3), and to be disheartened about the detrimental effects of our diet (Chapters 1 and 2). It is also easy to feel overwhelmed by the corruption and harm that politics too often play in our dietary recommendations (Chapter 2) and by the damage plastics are doing to our planet (Chapter 4). But it is also important to highlight that there is hope for a brighter future (as shown in the current chapter), so we do not fall into despair, which can lead to inaction.

As Jane Goodall has said, "The greatest danger to our future is apathy," and I wholeheartedly agree. So instead of apathy, anger, or sadness, let's now take some time to honor the hope that genuinely exists. Every individual has a role to play, and every individual can make a difference.

Example 1 Sheldrick Wildlife Trust

The Sheldrick Wildlife Trust (SWT), founded in 1977, is one of Africa's oldest wildlife charities and a leading conservation organization centered in Nairobi, Kenya.^{§§§§§§§§§§§§§§§} The SWT works closely with the Kenya Wildlife Service and the Kenya Forest Service to protect and save animal lives and to preserve habitats for the future of all wild species. ^{***************} The SWT engages in antipoaching initiatives and works to safeguard the natural environment, enhance community awareness, address issues related to animal welfare, and provide veterinary assistance to animals in need. The SWT also rescues and rehabilitates elephant and rhinoceros orphans, among other species.[313]

The SWT is most widely known for is its orphanage, situated in the Tsavo East National Park. The SWT has an entire team of rangers and keepers who rescue abandoned, lost, or injured animals who have all too often witnessed and suffered through the deaths of their own family members. These orphans are brought back to the orphanage, where their injuries – both physical and emotional – are treated and rehabilitated.

Once the orphans are old and experienced enough, they are reintegrated into herds that are protected within the park, which are often filled with prior orphans, their offspring, and some wild-living individuals. The amazing work of the SWT and the individuals who work with it in carrying out its mission and vision gives these orphans a second chance at a full life where they otherwise would not have survived.

To see the type of work that this organization engages in, visit their website or social media feeds and watch some of the videos they share of the orphans in their care. You will clearly see the bonds of love among the survivors and feel the hope that comes from these orphans having received a new lease of life. I often watch these videos and wish I could spend a bit of time myself caring for these beautiful elephants.

To learn even more about the SWT, I recommend the following wonderful book by Daphne Sheldrick: *An African Love Story: Love, Life and Elephants.* The book details her experiences of learning how to care for the orphaned elephants and also provides a deeper understanding of the losses

---

^{§§§§§§§§§§§§§§} www.sheldrickwildlifetrust.org/about
^{***************} www.sheldrickwildlifetrust.org/about/mission-history

that the elephant orphans experience – as well as their subsequent gains with the SWT.

A true testament to Mrs. Sheldrick, her family, and the SWT is that many of the elephant orphans who have been rehabilitated and subsequently reintegrated into the wild have since formed their own herds, and on occasion they visit the orphanage with their own offspring to introduce them to their prior protectors and keepers.

As of 2020, the SWT has hand-raised over 200 infant elephant orphans, and they have reintegrated over 100 orphaned elephants back into the wild herds of Tsavo East National Park. As of this writing, thirty-seven calves have been born to previously orphaned elephants who are now living in the wild, and there may be even more.[313],††††††††††††††††

In addition to the care that it provides to orphans, the SWT also operates antipoaching units in partnership with the Kenya Wildlife Service, and together they have made more than 2,800 arrests and removed more than 140,000 snares.[314],‡‡‡‡‡‡‡‡‡‡‡‡‡‡‡ The SWT has a canine unit with three Belgian Malinois dogs who have been trained to track and detect illegal wildlife "commodities" such as ivory, rhino horn, and bushmeat.[315] The SWT also has aerial surveillance teams, which are used in search-and-rescue operations and patrols and provide veterinary interventions for injured elephants and other wildlife.[314]

Due to recurrent droughts and limited rainfall in the Tsavo Conservation Area, especially during dry seasons, the SWT has built fifteen boreholes and windmills to improve the land's productivity, save habitat, and provide water to the wildlife that needs it.[316] All of these efforts have saved countless elephants, rhinoceros, and many other species.

The SWT offers a digital fostering program in which you can receive information about a specific elephant orphan. I have a foster elephant and it is really wonderful to see his progress. The SWT allows entrance to the orphanage for one hour each day (so as not to disturb the rhythms within the orphanage), and it is definitely on my own personal "bucket list" of places I would like to visit.

I thank them for their tireless efforts and for the hope they bring.

---

†††††††††††††††† There is a good chance that orphaned males may have impregnated wild females, whose calves are not counted in this figure (according to the SWT website).
‡‡‡‡‡‡‡‡‡‡‡‡‡‡‡ Snares are often used to capture wildlife – they are a very painful and slow method of killing.

Example 2 Sea Shepherd Conservation Society

What the SWT does for land animals in Kenya, the Sea Shepherd Conservation Society does for sea animals and the oceans. Sea Shepherd was formed in 1977 (coincidentally the same year as the SWT was formed) by Paul Watson, whose mission is to protect the world's marine life from "the destructive habits and the voracious appetites of humankind," including from illegal whaling ships and other groups that endanger wildlife.[317]

Although Paul Watson is often seen as a controversial figure to those who seek riches from the oceans – such as commercial fisherman, shark poachers, seal hunters, and whalers – Sea Shepherd has done much to protect wildlife around the world.

Sea Shepherd is a nongovernmental, nonprofit environmental organization operated primarily by volunteers who engage in both conventional protests and direct action to protect marine wildlife.[318] Some of their direct actions have included intervening against fishing and poaching in the South Pacific, the Mediterranean, and near the Galapagos Islands.§§§§§§§§§§§§§§§ Sea Shepherd also works to reduce ocean plastic pollution by organizing onshore cleanups near oceans, streams, and rivers and by removing discarded, lost, or illegal fishing nets from the oceans.[319]

Sea Shepherd volunteers have traveled to the Sea of Cortez, Mexico, to remove fishing nets and other plastic debris from the waters, which are known to kill the totoaba and the vaquita porpoise (Chapter 5).[277] Sea Shepherd volunteers also travel to Taiji, Japan, every year to document the capture and slaughter of dolphins in an effort to bring awareness of and an end to the practice. They similarly travel to the Faroe Islands to document the "Grindadráp," also known as "the grind," which is the dolphin and pilot whale drive hunting practiced there, in an effort to create peer pressure on people to stop these events.[320]

Sea Shepherd aims to protect the planet from the worst of humanity. Sometimes they succeed, sometimes they do not, but understanding their goals and their mission is key to convincing others to join in the journey to protect Earth.

I recommend both reading *Captain Paul Watson: Interview with a Pirate* and watching the documentary *Watson*. Both give much insight into why Sea Shepherd exists, what its goals are, and how we can all be better stewards of the oceans and wildlife. We all depend on the oceans for oxygen and food. This organization and its volunteers diligently act to

---

§§§§§§§§§§§§§§§ Much of this is detailed in the very recent 2020 documentary *Watson*.

protect the future of the oceans and demonstrate the importance of doing so.[321]

A key phrase from Paul Watson that I always revisit in my quest to educate others and encourage them to join in protecting the environment and the oceans is this: "If the ocean dies, we all die."******************* I am not saying we all need to go out and volunteer to be Sea Shepherds today, but I am suggesting that we all take decisive action now and do our part to improve our own stewardship as well as preserve and protect our one and only home, Earth.

Example 3 Firefighters from around the World Coming Together to Fight the Australian Bushfires

Not all heroes wear capes; many wear firefighters' gear.

In the 2019–2020 fire season in New South Wales, Australia – the worst fire season in Australia's history – 46 million acres (72,000 square miles) and 53 percent of the Gondwana world heritage rain forests in Queensland were burned.[322] Hot temperatures, drought, and high winds helped spread this tragic wildfire, and more than 1 billion animals are estimated to have been killed.[323],†††††††††††††††† As a result of these bushfires, over one-third of the world's koala population died, further increasing the extinction risk faced by this species.[324],‡‡‡‡‡‡‡‡‡‡‡‡‡‡‡‡ And all of this happened in just *one* fire season.[325]

Tragically, a similar scenario has occurred along the west coast and some western states of the United States during the summer and fall of 2020. Massive fires burned down several million acres, destroying property, forests, and the habitats of millions of animals up and down the coast.[326]

Firefighters from New Zealand and the United States traveled to Australia to help fight the fires together as comrades.[327] For at least twenty years now, firefighters from the United States and Australia have come together, traveling back and forth, to combat fires side by side.[328]

******************* Captain Paul Watson (www.ecowatch.com/paul-watson-if-the-ocean-dies-we-die-1882105818.html).
†††††††††††††††† Animals killed included mammals, birds, and reptiles.
‡‡‡‡‡‡‡‡‡‡‡‡‡‡‡‡ Woman saving Koala from brushfire: https://youtu.be/3x8JXQ6RTIU.

This spirit and desire to work together for the common good – admittedly attached to a specific time and place – needs to be replicated on a grander scale so that we can all participate in protecting Earth's future, and our own futures on it.

There is only one planet we know of with such a diversity of complex life, with an atmosphere and oceans that provide the water, oxygen, plants, and the building blocks of life for us all. There is no other planet or place we can live on, except in science fiction – and this is called *fiction* for a reason. So we all need to do our part, like those firefighters, to protect our one and only home, which is – quite literally – burning more often, more intensely, in more places, and for longer periods of time each year.

When the Australian outback was on fire, humanity showed the best of itself. Individuals were selfless, they were brave, and they fought for life. When our house is on fire, we must do all we can to save it, and right now, "our house *is* on fire."§§§§§§§§§§§§§§§§

Example 4 Reduced Emissions as a Result of the COVID-19 Pandemic

During the COVID-19 pandemic, we have all seen examples of individuals performing selfless acts of kindness and doing "good" to protect themselves and each other. Tragically, we have also seen examples of the opposite, where people act in ways that are selfish or self-serving.

However, my goal in using COVID-19 as an example in this "people doing good" chapter is that it demonstrates that even in the absence of government leadership and government action, we as individuals *can* make a difference – *must* make a difference – and that when all of our individual actions add up, they can make a huge, measurable difference.

I believe that there are lessons to be learned from the COVID-19 pandemic. One such lesson is that individual actions matter. During the COVID-19 pandemic, individuals were called upon to protect each other by keeping physical distance from each other to slow down the spread of disease. There is evidence that this strategy worked, especially in China, South Korea, Singapore, and New Zealand.[329] As a "silver lining," since many of us stayed in our own homes and didn't travel to work (or anywhere else), there was a reduction in greenhouse gas emissions from

---

§§§§§§§§§§§§§§§§ Quote from Greta Thunberg.

all transport and energy sources, especially in cities, which demonstrates that it *is* possible for our individual actions to reduce greenhouse gas emissions, to the benefit of us all.[7,330]

Of course, there were also many detrimental effects associated with the lockdowns during the COVID-19 pandemic, including economic downturns and negative effects on social and psychological well-being.[331] But nonetheless, there are lessons we can learn from some of these experiences that can be translated onto a wider stage, and hopefully without the negative impacts that the lockdowns had on psychological well-being.

The Paris Climate Agreement, ratified by 197 Member States (though only 189 have become party to it), sets to limit global temperature increase to 2°C (3.6°F) above preindustrial levels by 2030. In order to accomplish this goal, it is estimated that emissions need to fall below 25 gigatons per year by 2030. This is a tall order considering we currently emit more than this amount every year, and greenhouse gas emissions *rose* by more than 2 percent in 2018 as compared to 2016 levels, which is clearly a step in the wrong direction.[332]

It is suggested that between 2020 and 2030, annual global greenhouse gas emissions need to be reduced by 7.6 percent every year to come close to meeting this target.[333] This *can* be done by reducing carbon dioxide emissions by 20 percent, increasing the renewable energy market share to 20 percent, improving energy efficiency by 20 percent, ****************** engaging in more regenerative agriculture that sequesters carbon, and eating a more plant-based diet (more on these last two points in Recipe 1 of Part 2 of this book).[334] As daunting as these numbers seem, the lockdowns instituted during the COVID-19 pandemic demonstrate that some of these *can* be met by performing some of these aforementioned actions. In some locations, greenhouse gas emissions dropped by 25 percent or more during the lockdowns as a result of the following:

- Traveling less for work and allowing more work-from-home and telecommuting setups.
- More online meetings instead of in-person meetings. This especially makes a difference when in-person meetings require emissions-heavy flights.
- An increased number of safe zones and street closures to cars (and other petrol- or coal-dependent forms of energy transport) to be used instead for transport by bikes.

****************** Otherwise known as the 20/20/20 rule.

- Some cities and countries have encouraged this further by temporarily or permanently shutting down roads and transforming them into safe zones for foot or bicycle traffic only.[335,336,337,338,339]

The pace at which these shifts were made was rapid, demonstrating that changes *can be made*, and they can be made *quickly*.

While there were hits to the economy and to emotional well-being during the pandemic lockdowns, there are important takeaways from these experiences that can be translated in positive ways rather than in reactionary ways to help the economy, the environment, and our emotional well-being.[7] Some of these key takeaways include allowing more working from home where possible, conducting more virtual meetings, and instituting more no-automobile zones to encourage more foot and bicycle traffic, which also benefits cardiovascular and aerobic fitness and health.

Another important lesson learned from the COVID-19 pandemic is that food supply chains are at risk; they can break, and when they do, alternatives need to be available rather than adding pressure to an already fragile system, which is what happened in the United States.

During the pandemic, meat processing plants became super-spreader locations. As such, when they shut down to halt the spread, meat supply, at least in the United States, became more limited. This created quite a bit of panic among *some* shoppers.[340] On the other hand, other shoppers became increasingly interested in trying some of the many meat alternatives and plant-based analogs that were already on the market.[341] This is great to see, as plant-based alternatives can protect human health against chronic disease and also protect the environment, animal welfare, and vulnerable workers who were being forced to work in suboptimal conditions during the pandemic.[342]

These are just two small lessons that we can learn from the COVID-19 crisis that can help us to reduce greenhouse gas emissions and improve human health as we return to a new normal.[334]

Example 5 4ocean, a Nonprofit Ocean Cleanup and Advocacy Organization

While there are many individuals and organizations volunteering to help clean up and remove plastic from the oceans, the organization 4ocean especially caught my eye due to their campaigning strategy. 4ocean was founded by two men who grew up on the Florida coast. They traveled to Bali, Indonesia, for a surf trip and found a beach that was completely covered in plastic.[343] This organization is a nonprofit and does not accept donations; rather, it is funded by 4ocean product purchases, many of which are made from the repurposed trash or plastic they find on beaches and in the ocean. The organization also hosts community cleanup days that you can get involved in.

Other organizations attempting to protect the oceans include Sea Shepherd, Trash Free Seas, the Sea Change Project, the Surfrider Foundation, Ocean Conservancy, the 5 Gyres Institute, Oceana, Bye Bye Plastic Bags, The Nature Conservancy, the Lonely Whale Foundation, the Bahamas Plastic Movement, and Parley for the Oceans.[††††††††††††††††] If you are looking for organizations to work with, learn from, buy from, or donate to, I strongly recommend looking at their websites.

Example 6 Akashinga[‡‡‡‡‡‡‡‡‡‡‡‡‡‡‡‡‡]

Akashinga is a part of the International Anti-Poaching Foundation (IAPF). The IAPF was founded in 2009 by Damien Mander, an Australian special-ops sniper and war veteran turned antipoaching hero – as a nonprofit organization dedicated to conservation and antipoaching in Africa.[344]

In 2017, Mander began the Akashinga program, which I highlight for two reasons. The first is that it is a nonprofit organization dedicated to fighting for wildlife, to helping to end poaching, and to protecting and defending nature. The second and perhaps even more important reason is that Akashinga is a woman-led, woman-run, woman-only unit of the IAPF. Akashinga uses community-oriented approaches to end poaching and was developed as a way to embolden and empower victims of sexual assault or domestic violence, women who are single mothers or abandoned wives, or women who are AIDS orphans.[344]

So far, Akashinga has lowered elephant poaching by 80 percent in Zimbabwe's lower Zambezi Valley, an amazing testament to this organization. Akashinga hopes to recruit up to 1,000 women to be female rangers by

---

[††††††††††††††††] For more information on all of these wonderful organizations, trying to do good for the planet and the oceans, see www.southernliving.com/travel/organizations-fighting-to-save-oceans.
[‡‡‡‡‡‡‡‡‡‡‡‡‡‡‡‡‡] www.iapf.org/news/akashinga

2025. By training and empowering women, many benefits are bestowed on the local communities, including improved educational opportunities, personal development, and better standards of living.[344] To learn more about this beautiful organization and the good work they do, I urge you to watch the fourteen-minute video on their website.[SSSSSSSSSSSSSSSSSS]

Example 7 Animal Rescues and Sanctuaries

There are a number of organizations throughout the world that are doing so much good. Many of these organizations take in animals that have been abandoned or who have been abused either as performing animals or working animals. Some of these organizations rescue animals, rehabilitate them, and release them to the wild if they are able to do so, while others strive to give their rescued animals who are unable to be released back into the wild the most natural and free life that they can. I will name a few examples of animal rescues and sanctuaries, but there are many more out there worth exploring and supporting.

Tikki Hywood Foundation and the Pangolin Men: This nonprofit organization in Harare, Zimbabwe, dedicates its mission to saving, rehabilitating, and releasing (where possible) endangered and lesser-known animals such as pangolins.[********************] The Tikki Hywood Foundation was founded in 1994 as a nongovernmental organization (NGO) that operates as a twenty-four-hour wildlife rescue. This organization also works on legislation to uphold laws that protect fauna and flora in Africa.[345] They also work with local communities to assess and mitigate wildlife and domestic animal conflict with neighboring human settlements. Education and raising awareness to change perceptions, beliefs, and habits (e.g. through animal welfare workshops and education in primary schools) are other key missions of theirs.[345] It is well worth viewing their online video "Pangolin Men Saving the World's Most Trafficked Mammal."[††††††††††††††††††]

Performing Animal Welfare Society: The Performing Animal Welfare Society (PAWS) is based in Northern California. PAWS was founded in

---

[SSSSSSSSSSSSSSSSSS] www.iapf.org/the-film

[********************] For more information, see www.tikkihywoodfoundation.org.

[††††††††††††††††††] https://youtu.be/ypXBlyeUHdE

1984 and has rescued and provided a humane sanctuary for animals who are victims of the exotic animal trade or were performing animals.‡‡‡‡‡‡‡‡‡‡‡‡‡‡‡‡‡‡ PAWS also investigates reports of abused performing and exotic animals and assists in investigations and prosecutions by regulatory agencies.[346] PAWS has rescued and rehabilitated African elephants from zoos, most of which were captured from elephant cull operations in Africa and have experienced traumatizing events in their lives. PAWS has also rescued and rehabilitated Asian elephants from circuses, which often use harsh and punishing training techniques.[346]

PAWS has also rescued tigers from roadside zoos or attractions, where cubs are often bred for photo shoots or for the exotic pet trade. These cubs are often forcibly removed from their mothers soon after birth so that they can be bottle-fed and handled by people, leaving them vulnerable to deadly infections. Eventually, these cubs grow too big to be handled and are put back into the breeding population to create more cubs. Many of these tigers have had numerous traumatizing experiences. PAWS gives them the space to live more like a tiger should (as they are unable to be reintroduced into the wild).[346]

PAWS works with all kinds of animals, with other examples being monkeys kept as exotic pets, bears kept as tourist attractions, and lions abused as performing circus animals who are unable to be released back into the wild.[346],§§§§§§§§§§§§§§§§§§§§

Dolphin Project: This is another nonprofit organization, which was founded in 1970 and is dedicated to the welfare and protection of dolphins worldwide.[347] Their mission is to educate the public about dolphin captivity, to end dolphin exploitation and slaughter, and, where feasible, to rehabilitate and release captive dolphins back into the wild.[347] Dolphin Project has helped many dolphins around the world over the past fifty years. They brought attention to the brutal drive hunts that take place along the coast of Taiji, Japan (which I mentioned in Chapter 6) and have investigated and advocated for economic alternatives to dolphin slaughter, as can be seen in the film *The Cove*, which features Dolphin Project's founder.[305]

Dolphin Project is the longest-running anticaptivity dolphin welfare organization in the world. Dolphin Project has developed protocols that

---

‡‡‡‡‡‡‡‡‡‡‡‡‡‡‡‡‡‡ PAWS is a charter member of the Global Federation of Animal Sanctuaries, formed in 2007 by "nationally and globally recognized leaders in the animal protection field for the sole purpose of strengthening and supporting the work of animal sanctuaries worldwide."

§§§§§§§§§§§§§§§§§§§ For the most part, these animals cannot be released back into the wild as they have retained very little, if any, natural instinct to survive in the wild.

are used to "untrain" captive dolphins, allowing for their eventual release back into the wild. They have successfully released twenty dolphins back to the wild.[348]

Borneo Orangutan Rescue: International Animal Rescue's Borneo Orangutan Rescue works primarily in Kalimantan, the Indonesian part of the island of Borneo, to rescue and care for baby orangutans who have been taken from their mothers to be illegally sold as pets. They also help adult orangutans who have spent their entire lives in captivity, chained up or imprisoned in tiny cages. Borneo Orangutan Rescue also has a team that is dedicated specifically to rescuing orangutans who have been left stranded due to human–orangutan conflict or palm oil production that has destroyed their habitat. The rescued orangutans are relocated to safe areas of protected forest, or if they are no longer able to survive in the wild, they are given a permanent home at the rescue center.[349]

Example 8   Lewis Pugh Foundation

Finally, the Lewis Pugh Foundation is an organization that works toward the preservation and conservation of oceans by working with NGOs, governments, scientific institutions, and concerned citizens around the world to create protections for over 2.2 million square kilometers of ocean.[350] They currently have three campaigns in progress to protect the oceans. These include the 30 × 30 campaign, the Antarctica 2020 campaign, and the Arctic Decade campaign.

The *30 × 30 campaign* is an extremely ambitious campaign with the goal of protecting 30 percent of the world's oceans by 2030. To call attention to this campaign, Lewis Pugh swam the full length of the English Channel from Land's End to Dover, a full 528 km (330 miles), which took 49 days. As a result, the British government became the first to call for protecting 30 percent of the world's oceans by 2030, with other nations hopefully joining in as well.[350]

The *Antarctica 2020 campaign* began in December 2016 following the establishment of the Ross Sea Marine Protected Area. The Lewis Pugh Foundation was instrumental in getting 2 million square kilometers (772,200 square miles) of the Southern Ocean protected.[350]

Finally, the *Arctic Decade campaign* demonstrated the effects that climate change is having on the Arctic. The swim that Lewis Pugh completed for this campaign was so poignant because, in years past, it could not have been completed, as the North Pole was previously covered in frozen ocean. However, in the summer of 2007 it was not, and Lewis Pugh became the first person to swim across the North Pole.[350]

In each of his campaigns – and there have been several more – Lewis Pugh calls attention to important issues surrounding the ocean through extraordinary actions and feats of courage. He is someone who leads by example and demonstrates the courage that is needed to produce actionable change.

Clearly, there are hundreds of organizations doing truly wonderful work, and I encourage you to search them out and find your own examples of organizations that are doing good. Knowing all of these organizations exist does give me hope for the future; I hope that it also instills hope in you.

Lastly, here are some case examples showing how organizations have helped curtail wildlife exploitation by working closely with individuals, giving them the tools and assistance they need to transform their lives and livelihoods to be protective of the environment and species rather than exploitative.

Case Examples

Often individuals who engage in wildlife poaching and trade feel they have no other choice.[351] These individuals often have insecure livelihoods and food supplies with few tangible assets or resources to turn to.[352] Most individuals who participate in wildlife poaching would prefer options that are more sustainable in the long term to provide for their families. Most of them understand that when they engage in poaching and the degradation of wild ecosystems, they cause irreparable harm to those species and to their own personal sources of income or livelihood.[353,354]

The problem is that drought, excessive heat, flooding, environmental degradation, and human–animal conflict decrease the productivity and habitability of the land and increase the risk that individuals and households will experience negative health, nutritional, or livelihood outcomes, forcing them into coping strategies they do not want to engage in, such as wildlife poaching or trade.[18] This is a most vicious cycle.

Sometimes there are other livelihood strategies or options available that are both safer and more sustainable, including sourcing wild foods or getting paid to do labor for someone else. Other options may include the use of safety net programs******************** if they exist, selling other personal assets including livestock or agricultural products, or obtaining small transfers and loans where possible.[355] Unfortunately, though, sometimes these strategies, or even combinations of these strategies, simply are not enough.

In order to seek viable alternatives to poaching, buy-in and acceptance of these alternatives *by poachers* is key. If someone's source of income is going to be taken from them, there needs to be a better option available to replace it.

Below are some case examples that demonstrate how changing a livelihood from poaching or trading in endangered species to one that focuses on ecotourism, conservation, and positive wildlife experiences can benefit species, their habitat, and the local economy altogether, providing long-term and more stable economic and social benefits for the individuals involved.

> *Case 1:* A film that does a wonderful job of demonstrating these issues is *Racing Extinction*, which documents the mass killing of manta rays in the Maldives, Indonesia, and in the Philippines for their gill plates, which are traded in Asian markets for medicinal purposes.[356,357] Manta rays are filter feeders that thrive on zooplankton and can grow to be more than 2 meters (7 feet) wide.†††††††††††††††††††††† Manta rays were on the verge of local extinction in these waters due to poaching. Yet when the trade in manta rays was replaced with ecotourism and "swim-with-manta" programs, local incomes significantly *increased* as more tourists came and spent money in these communities. Moreover, these communities earned prestige and more positive reputations for wildlife conservation, habitat rebuilding, and species protections.[356] In fact, one study looked at the overall cost–benefit analysis of these changes and revealed that live manta rays are worth twenty-eight times more than dead ones in terms of ecosystem and livelihood benefits.[358] Additionally, working with local organizations to keep manta rays and their habitats thriving ensures that this is a sustainable, long-term livelihood strategy that benefits all.

******************** Safety net programs may include cash transfers (allowances) to families or food transfers (vouchers/direct food provisions) to families (for examples, see www.jstor.org/stable/23339844?seq=1#metadata_info_tab_contents).

†††††††††††††††††††††† Zooplankton are a type of heterotrophic plankton that range from microscopic organisms to species that grow larger with age, representing the larval or young stages of animals (e.g. crabs, lobsters, fish; https://biologydictionary.net/zooplankton).

Case 2: Similarly, exchanging elephant poaching for ecotourism within elephant habitats in Africa helped local communities earn seventy-six times more income than they earned when they were poaching and selling the ivory of dead elephants.[359] Again, by protecting the elephants' habitat, the ecosystem resources, and the elephants themselves, the local peoples ensure a long-term commitment to the improvement of their communities.

Case 3: The film Racing Extinction also highlights how protecting sharks and their habitats is of critical importance. In areas where sharks are now protected (rather than hunted and finned) and where ecotourism is encouraged, entire ecosystems have been revived and are now thriving along with local businesses and individuals' livelhoods.[356]

Case 4: In yet another wonderful example, a former sixth-generation dolphin hunter named Ishii, from Futo, Japan, changed his livelihood from hunting dolphins into a successful income-generating business involving dolphin watching, protection, and advocacy.‡‡‡‡‡‡‡‡‡‡‡‡‡‡‡‡‡‡‡

Even more examples of stories with similar hopeful and positive outcomes can be found all around the world, demonstrating hope and that it is possible to change for the better, to protect wildlife, and to improve local communities' livelihoods and well-being all at the same time.

These case studies show how intelligent, sustainable ecotourism and the conservation of nature can improve livelihoods, increase livelihood security, bring in more money, bring more attention to the topic, and become self-perpetuating businesses. The basic tenet is this: "The more animals, the more business."[360]

Some of these examples effectively show how protecting even *one* keystone species can *also* protect other species and the habitats they all share. Moreover, protecting species and engaging in sustainable ecotourism can promote the stewardship of wildlife and habitats while also saving and improving the lives of the animals and the local communities who support and depend on these species.[361]

Shifting livelihoods and behaviors shows a commitment and an understanding that losing even one species represents more than simply losing its individuals. When a species goes extinct, the world loses more than *just* a

---

‡‡‡‡‡‡‡‡‡‡‡‡‡‡‡‡‡‡‡ Ishii's story can be viewed on YouTube at www.youtube.com/watch?v=V_lEa68efps#action=share.

unique genetic code. The world also loses the benefits that this species provided within its own ecosystem and in the web of life among species. Losing a species – any species, even one that appears insignificant to us – *will* affect us, and in some way *will* lower our chances for survival as a species on *this* planet.[278]

Protecting biological diversity is not simply about protecting what is pretty or admiring only what we believe deserves protection. Protecting biological diversity ensures a healthy planet, decreases the risk of blight, and decreases the risk of the spread of disease and environmental collapse.[278]

Monocultures (as discussed in Chapter 1), where one species of crop is grown or one type of livestock is reared, have an increased risk of failure of the entire crop or flock.[362] If a blight comes through the field, hundreds of acres of crop can be lost. If an illness is carried through a herd, millions of dollars-worth of lives could vanish. Protecting biological diversity and working toward sustainability at local, regional, and global scales are crucial for the health and safety of our planet and the health and safety of us all.[278]

E. O. Wilson, the famous American biologist, naturalist, writer, and professor, said it best: "Animals [and other species] do not need us to survive, but we sure do need them."

Each animal and plant has its place in the web of life. When we disrupt that web, innumerable and immeasurable changes occur that may have a profound effect on the strength, flexibility, and adaptability of that web. All life is the product of evolution, and species survive through the interdependent role each plays in this web of life. It is imperative that we as individuals protect that web as much as is humanly possible – not just for its own sake, but for our own future survival as well.

Part 2, which begins in the next chapter, provides practical, everyday ways in which we can empower ourselves to be better stewards of the planet, to be good, and to do good.

8

PART 2

Protecting Earth, One Recipe at a Time – An Introduction

> You cannot get through a single day without having an impact on the world around you. What you do makes a difference and you have to decide what kind of a difference you want to make.
> —Dr. Jane Goodall

In Part 1 of this book, I explained how close we are to approaching a tipping point where, no matter what we do, Earth will continue to warm at a runaway pace, species will perish at increasing rates, and the world will be less hospitable to us, to crops, and to all life.

While no one knows exactly *when* we will hit this tipping point, we *do* know that every day we are getting closer to it (and perhaps we have already hit it).

It seems that not enough governmental organizations around the world are taking the immediate, decisive, or direct actions that are needed to tackle climate change. It also seems that not enough governmental organizations are considering the problems outlined in this book seriously enough. But, then again, not enough governmental organizations took the COVID-19 pandemic seriously enough either.

So what are we to do in the absence of governmental action?

Given the urgency of the global situation we find ourselves in, we must take it upon ourselves to do what we can, when we can, to make whatever positive difference we can to protect ourselves and our home.

"We are the last generation that can stop climate change."[363] So what are we waiting for?

In Part 2 of this book, I give you twenty-one ideas, pieces of knowledge, tools, or "recipes" that can empower you and others, perhaps even national leaders, on how to make impactful changes that can simultaneously improve *your own health* and help stop or even reverse climate change.

The pandemic has given us the chance to see how important individual efforts actually are. We have seen that when individuals take it upon themselves to follow health recommendations – especially ones that are low effort such as using personal protective equipment including cloth facemasks to protect themselves and others – overall death rates can be significantly lowered and the effects of the virus reduced.

In contrast, we have also seen several examples where individuals and government leaders ignored health experts' recommendations. Where this occurred, outbreaks lasted far longer, became far more widespread, many more people lost their health and/or their lives, and economic recovery took significantly longer, if it even began.

The issues surrounding climate change are very much in this same vein. There are countries that take climate change *very seriously* and actively work toward improving the environment – by planting trees, protecting habitats, and lowering their emissions. But then there are countries that refuse to even acknowledge that climate change is real, that deny the science and instead promote the use of fossil fuels, that roll back environmental regulations, and that endanger their populations and the world. Where this is happening, our only real choice is to take action and do what *we* can as individuals.

If we choose to do nothing, we will continue on Earth's warming and deadly trajectory. However, if we choose to act – to do something, be it any one (or more) of the ideas or "recipes" addressed below – we stand a better chance of protecting the planet and ourselves.

The choice is yours; the choice is mine.

Throughout Part 2 of this book, I talk about many individual actions that we can take to make a difference to our health and to the environment. In discussing each action, I explain why it is important and what the research says about it, and I give a few tips or resources regarding how to make it doable and achievable.

In isolation, each idea discussed below may seem small, but taken together, these small ideas can add up to big, significant benefits to your health and to the environment.

Change starts with me; change starts with you.

Any time you doubt what one person can do, look back at the COVID-19 pandemic and realize that one individual can make a difference to another individual, with knock-on effects that spread from one person to another, and to another still, for good or for bad, just like how a rock thrown into a placid pond produces ripples that grow outward, carrying energy and momentum.

> Never doubt that a small group of thoughtful, committed citizens can change the world; indeed, it's the only thing that ever has. —Margaret Mead

You can be one of those thoughtful, committed citizens that helps to change the world with your actions – one person at a time, one action at a time, and one "recipe" at a time.

We are more powerful than we may realize, both with our actions and with our votes. When we vote, we vote to enforce specific policy initiatives as well as social and political change. When we buy goods and services, we are voting with our wallets. What we choose to invest in – socially, politically, or fiscally – can make a huge difference. It starts with each individual, but collectively our actions can be very powerful.

When decisive and benevolent leadership is not there, it is up to each of us to make a difference. We are *not* powerless to change things. We can be part of social or political movements that aggregate our efforts to nudge or, better yet, shift policy and politics toward improving health, sustainability, and conservation.

We can start today – we can start in this moment – to protect our health against chronic disease, to protect populations against food insecurity, and to protect Earth against pollution, species extinctions, and climate change.

You are about to dive into twenty-one ideas, each providing suggestions on "how to" make it achievable. The last idea gives you a number of documentary and reading resources so that if you want to dive deeper into any of these topics you know where to look.

Finally, at the end of this Part 2, you will find "recipe" cards for each idea that you can cut out and tape to your wall as a reminder or as a little cheat sheet for how you can make a positive difference.

Each idea is meant to be a small beacon of *hope* that, through action, education, and role modeling, more people will also become eager to participate and engage in healthy behaviors, sustainability, and environmental protection. It is my hope that our many small changes will add up to large, easy-to-measure changes that give us healthier and longer lives as well as a more habitable Earth.

Working toward improving our health and protecting the environment are *not* mutually exclusive. We can have healthier lives and a healthier planet at the same time – it is not one or the other. We can eat more healthfully, feed more people, and protect the environment all at the same time. In fact, that is what we *need* to do, and we need to do it now.

Although it may seem challenging, one thing that the COVID-19 pandemic has taught us is that change – even rapid change – *is* possible when the will is there. We have seen the creation of effective therapeutics and vaccines at record pace. We have seen companies and organizations work together for the common good. We can be inspired by these efforts and translate them for the purpose of slowing and stopping climate change, and we need to do so urgently.

There have been far too many climate meetings that look good on paper but, in reality, *do* nothing. It is time we take matters into our own, individual hands, and we must start today to protect both our individual health and the health of our home, Earth.

To guide you in this journey, here are twenty-one ideas or "recipes" in order of importance – at least as I see it – regarding how much of a difference they make to our health and to the overall health of the environment. These recipes are presented as follows:

Recipe 1   Eat More Plants (and Significantly Less Meat)!
Recipe 2   Buy Organic or Fair Trade
Recipe 3   Eat More Omegas (but Ditch the Fish)!
Recipe 4   Decrease the Amount of Food You Waste
Recipe 5   Buy in Bulk to Reduce Packaging
Recipe 6   Can, Jar, or Freeze
Recipe 7   Compost
Recipe 8   How to: Home and Community Gardens
Recipe 9   When You Cannot Grow It Yourself, Shop at Local Farmers Markets and Use Community-Supported Agriculture!
Recipe 10   Reduce, Reuse, and Refuse the Plastic: There Are Better Alternatives!

Recipe 1 Eat More Plants (and Significantly Less Meat)!

> Nothing will benefit health and increase the chance for survival on
> Earth as much as the evolution to a vegetarian diet.
> —Albert Einstein

One of the most effective and proactive things we can do to help the
environment and our *own health* is to significantly – and immediately –
reduce our intake of animal products and instead eat more plant foods,
including plant-based proteins, whole grains, nuts, seeds, pulses, vegetables, and fruits.[364]

As I mentioned in Chapter 1, at least one-quarter of all greenhouse gas emissions (and maybe as much as 50 percent) come from the
foods we eat and how they are produced. That's more greenhouse gas
emissions than are produced by the entire transportation sector (see
Figure 8.1).[34,§§§§§§§§§§§§§§§§§§§§§]

While it might be difficult to imagine how something as simple
as eating creates so much environmental damage, when you dive deeper
into the topic – as we did in Part 1 – it begins to make sense.

Every year, billions of animals are bred and raised for human
consumption. In Western countries such as the United States, Australia,
and Canada, over 55 billion individual animals are killed each year for

---

§§§§§§§§§§§§§§§§§§§ In Figure 8.1, "Agriculture, Forestry and Other Land Use" represents
greenhouse gas emissions from the food industry.

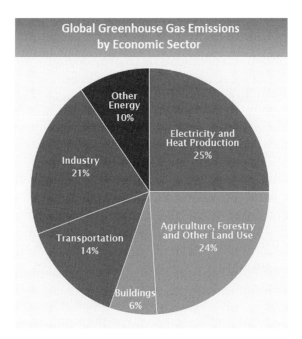

Figure 8.1 Global greenhouse gas emissions by economic sector
Image credit: Intergovernmental Panel on Climate Change, 2014 (www
.ipcc.ch/report/ar5/wg3)

this purpose. This turns out to be roughly 170 animals (80–115 kg or
175–250 pounds of meat) per person per year.********************

In contrast, in developing countries, the average person eats
around 25 percent as much meat as we do in the West (an average of
10–40 kg or 22–88 pounds per person per year). This discrepancy is not
because individuals in developing countries want to eat less meat; rather,
it is because historically they have not had the means, the income, or the
resources to buy as many animal products.[78],††††††††††††††††††††††

******************** https://animalclock.org

†††††††††††††††††††††† This presents some complicated issues that have a number of health,
environmental, and ethical and moral implications, including:

(1) *Worst-case scenario:* Those who have not had the opportunity to eat animal products in
abundance (namely in developing countries) should now be able to eat what they want
when they want to, if they can afford it. If populations in developed countries make no
changes to their own consumption patterns, this will have the most detrimental health and
environmental implications (this would be really bad).

Consider for a moment what would happen if all 7.8+ billion people on the planet ate as many animal products as we do in the West. If that happened, over 1.25 trillion (1,250,000,000,000 – just so you can see the number of zeros) animals would need to be raised and killed every year, and far more forest would need to be cut down. Already, too much forest is slashed each year. This is a recipe for disaster and is not sustainable.

These animals that humans eat require a lot of calories to grow and produce many waste products that are environmentally detrimental, including the greenhouse gases methane and nitrous oxide. Forests – which both produce oxygen and absorb carbon dioxide – are cut down at ever-increasing rates to create land for grazing and/or to produce crops for these billions of animals to feed on. Figure 8.2 shows a breakdown of where greenhouse gas emissions come from within the food sector.

The amount of resources used to raise animals for human consumption is not sustainable – especially in the setting of a growing human population and a growing desire for animal products (both meat and dairy).[78] The world's food supply will need to increase by 38 percent and meat production by 62 percent over the next 30 years.[78] More domesticated animals will need to be raised and more calories will need to be grown, likely at the expense of forests. Already 50 percent of all the world's ice-free land is used for agriculture. There simply is not enough new or additional arable land to achieve this growth.[363] Therefore, we will need to be more efficient with the land we use, and the easiest way to achieve this is to shift our diets to foods that use fewer resources to produce – plants.

(2) *Slightly better-case scenario:* Those of us in developed countries who have been eating more than our share of animal products and have contributed most to environmental degradation, pollution, and ill health should lead by example and change (reduce) our consumption habits to make room for those who have not yet had the opportunity to eat these foods. This would allow those who have not yet had as much of these foods to be able to consume them more, perhaps maintaining current levels of environmental degradation (this would also be not so great).

(3) *Moderate-case scenario:* Those of us in developed countries who have been eating far too many animal products for far too long significantly lower our intake while populations in developing countries only moderately increase their intake (this would likely maintain or slightly lower environmental degradation).

(4) *Best-case scenario:* Populations in developed countries significantly reduce their intake of animal products to levels at or below those of populations in developing countries, and populations in developing countries do not increase their intake of animal products but improve their nutrition through plant-based sources.

Global greenhouse gas emissions from food production

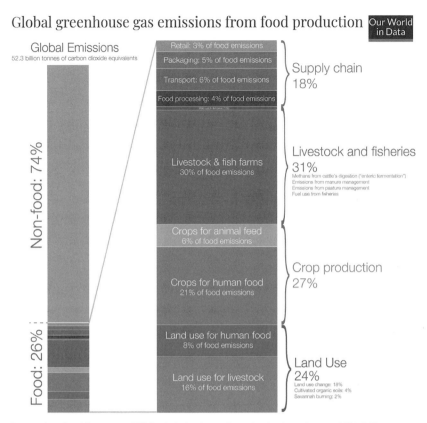

Figure 8.2 Where greenhouse gases come from within food production
Image credit: Ritchie (2019), Our World in Data (https://ourworldindata.org/uploads/2019/11/How-much-of-GHGs-come-from-food.png)

As you can see in Figure 8.3, as a general rule, raising animals for food requires far more resources (land and freshwater) and produces far more waste (greenhouse gases and excrement) – calorie for calorie – than growing plant-based foods.‡‡‡‡‡‡‡‡‡‡‡‡‡‡‡‡‡‡‡‡

‡‡‡‡‡‡‡‡‡‡‡‡‡‡‡‡‡‡‡‡ Per calorie, animal-based foods consume more resources and produce more greenhouse gases than plant-based foods. Land use: twenty-five times as much land for beef and three times as much land for pork compared with plant-based foods. Freshwater: approximately six times as much water for beef and two times as much water for farmed fish compared to plant-based foods. Greenhouse gas emissions: sixteen times as much for beef, four times as much for dairy, and two times as much for eggs, pork, and poultry compared to plant-based foods.

Animal-Based Foods Are More Resource-Intensive than Plant-Based Foods

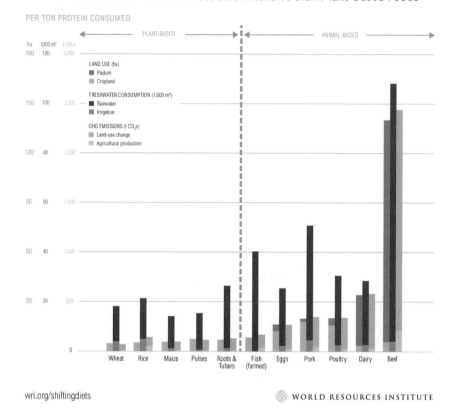

wri.org/shiftingdiets

✺ WORLD RESOURCES INSTITUTE

Figure 8.3 Amount of resources used in producing various types of foods. GHG = greenhouse gas
Image credit: World Resources Institute working paper, "Shifting Diets for a Sustainable Future" (www.wri.org/publication/shifting-diets)

In addition to using fewer resources, there is evidence that improving soil management and farming techniques will allow significantly more calories (crop volume) to be grown on the same amount of land (compared to chemically intensive agriculture or animal agriculture).§§§§§§§§§§§§§§§§§§§§§§§ Another benefit to improving soil

---

§§§§§§§§§§§§§§§§§§§§§§§ Chemically intensive agriculture is the type of "conventional" agriculture practiced in much of the Unites States. Large amounts of pesticides, herbicides, and fertilizer are used. Improved soil management consists of increased use of compost, having more trees and cover crops, better crop management and rotation, and also improved water management (to name a few).

management is that it also increases the amount of carbon that can be stored in the soil (this is called soil carbon sequestration).

Models show how soil carbon sequestration can slow down and possibly even reverse global warming by pulling carbon *from* the atmosphere.[365,366,367] Theoretically, the amount of carbon that can be stored in soil is more than the amount of carbon that humans release into the atmosphere each year from all sectors.[367]

Furthermore, having more carbon in the soil also increases soil fertility and enhances crop growth without the use of synthetic fertilizers. Increased crop growth then further increases carbon sequestration – this is a beautiful synergistic cycle. Think about it – it is possible to have more, safer, and less expensive food, fewer greenhouse gases, feed more people on less land, and be healthier, all through simple shifts in our diet and in how our foods are grown.

Plant-Based Diets and Better Health

There is a universal goal to live the healthiest life possible. Keeping this in mind, it is necessary that you understand that what you eat greatly affects your health and may be one of the most important things you do to be healthy, to prevent a range of chronic diseases, and to lower your healthcare bills.

The leading causes of death around the world are not infectious diseases. For the last couple of decades, the majority of global deaths have been from noncommunicable (noninfectious), nutrition-related chronic diseases (NCDs). NCDs are responsible for 70 percent of all deaths worldwide, or roughly 40 million people per year.[27,368,********************] And although more than 2.7 million individuals (after one year from when the COVID-19 pandemic was announced) have tragically died from COVID-19, these deaths are far outnumbered by the deaths associated with NCDs each year.

The fact is, many NCDs are preventable and reversible.

High intakes of meat, dairy products, processed foods, sugar, and salt and low intakes of whole grains, nuts, seeds, vegetables, and fruits increase the prevalence of NCDs (and premature death).[369] Sadly, for the first time in decades, life expectancy is declining in high-income

******************** Whenever I mention NCDs, I am referring to noninfectious diseases such as heart disease, stroke, pulmonary disease, diabetes, obesity, kidney disease, and cancer.

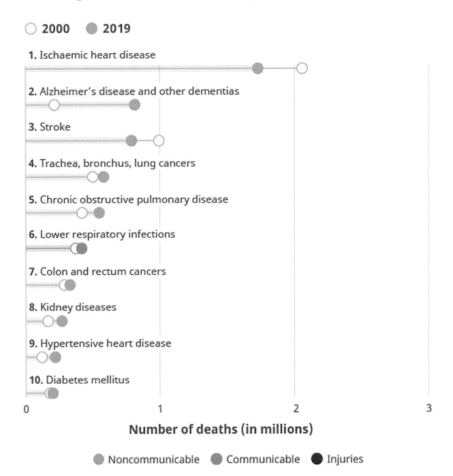

Figure 8.4 Top ten causes of death in 2000 and 2019 (high-income countries) Available at www.who.int/news-room/fact-sheets/detail/the-top-10-causes-of-death

countries due to the spread of diet-related diseases, and this is causing large cohorts of people to die younger than their own parents.

To see how the prevalence of NCDs has changed in the past nineteen years, see Figure 8.4, which shows the top ten causes of death for high-income countries such as the United States, those of the European Union, Australia, and Canada for the years 2000 and 2019.

Obesity, chronic disease, and premature death are huge public health crises that often are underaddressed, and while we may think diet-related diseases are limited to high-income countries, the fact is that the prevalence of chronic diseases has also been rising rapidly in developing – low-income – countries where populations are undergoing a transition from being primarily rural and farm-based to being more urban, sedentary, and reliant on processed foods and diets that are richer in animal products.[370]

Figure 8.5 shows the top ten causes of death for 2000 and 2019 in developing, low-income countries. Although infectious diseases such as lower respiratory tract infections, parasite- or vector-borne diseases, tuberculosis, and HIV are still in the top ten causes of death, Figure 8.5 also shows that NCDs including heart disease and stroke are also in the top ten, and they have actually increased in their prevalence in the past nineteen years (they are currently third and fourth, while in 2000 they were sixth and seventh). Unfortunately, they are continuing on this upward trajectory.[69]

More evidence of this phenomenon comes from China. When meat consumption in China quadrupled between 1970 and 2012, the proportion of deaths due to chronic disease rose from 40 to 74 percent (in that same time frame), and this proportion is still rising.[371,372,373]

The good news is that a plant-based diet reduces the risk of most NCDs. A plant-based diet also reduces the likelihood of developing obesity and has been shown to enhance quality of life by helping you be healthier and fitter and by adding years to your life.[374,375]

We who live in countries where there is too much food and too many calories available, such that they often end up as waste, do not need animal proteins to be healthy. In fact, populations around the world who eat a plant-based or Mediterranean-style diet actually tend to live longer and healthier lives than do populations where people eat more meat and dairy products.[130,376]

As mentioned throughout this book, many of the dietary habits associated with an increased prevalence of obesity and NCDs are also those that contribute to many of the environmental crises we face and to the origination and spread of new and emerging infectious diseases, such as COVID-19 (see Chapter 1 for more discussion on this).[377]

Given all of the connections among eating more animal products, poorer health, negative environmental effects, and the increased risk of emerging diseases, it is evident now more than ever that many of

Leading causes of death in low-income countries

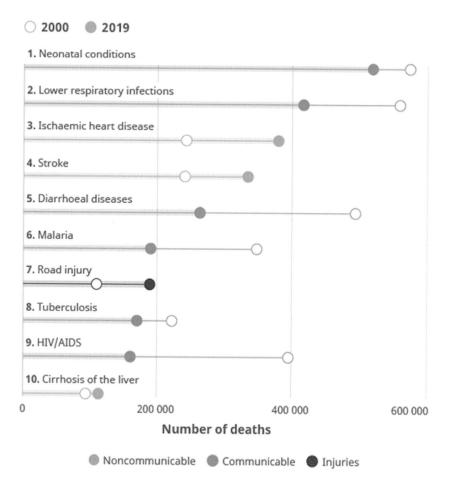

Figure 8.5 Top ten causes of death in 2000 and 2019 (low-income countries)
Images credit: World Health Organization, Global Health Observatory (GHO) data
(www.who.int/gho/mortality_burden_disease/causes_death/top_10/en)
Available at https://www.who.int/news-room/fact-sheets/detail/the-top-10-causes-of-death

us should shift our dietary habits to being more plant-based (with far fewer animal products). The benefits of making these changes will be far-reaching.

As I like to tell my students and patients, "What is good and beneficial for *our* health is good and beneficial for the *planet*, and likewise, what is good and beneficial for the *planet* is good and beneficial for *our own* health." This is because plant-based diets are associated with all of the following:[378]

- A significantly lower risk of obesity and many types of chronic disease;
- Increased longevity;
- Less inflammation, which can help the gastrointestinal tract, heart, and immune system;
- A lower body weight and body mass index, being heavier is associated with chronic disease risk;
- Lower blood pressure and decreased risk of hypertension, in part because plant-based diets are typically higher in potassium and magnesium and lower in salt and fat;[378]
- Lower hemoglobin-A1c levels – also called glycated hemoglobin – which measures how much glucose is attached to hemoglobin proteins in blood cells and is often used to diagnose individuals with diabetes;[†††††††††††††††††††††]
- Lower cholesterol levels and a significantly lower risk of future heart attack or stroke;[379]
- Lower risk for chronic kidney disease or kidney injury;
- Improved levels of emotional and physical well-being and self-efficacy and decreased levels of depression.[380]

On top of these benefits, plant-based diets are also naturally high in fiber, water, vitamins, minerals, antioxidants, and other anti-inflammatory compounds, and they are prescribed – by those who are in the know – to reverse many chronic diseases.[381]

Additionally, plant-based diets can slow down climate change and land degradation and prevent habitat losses, deforestation, and species extinctions.[44,382,383,384,‡‡‡‡‡‡‡‡‡‡‡‡‡‡‡‡‡‡‡‡‡] In a nutshell (pun intended), eating a plant-based diet is a win–win for our health, the environment, and for wildlife species that are going extinct at unprecedented rates due to competition and conflict with humans and deforestation.

[†††††††††††††††††††††] Hemoglobin-A1c is also used to monitor how well individuals with diabetes are controlling their blood glucose levels.
[‡‡‡‡‡‡‡‡‡‡‡‡‡‡‡‡‡‡‡‡] In Central and South America, forest cover has been reduced by almost 40 percent in the last 40 years due to deforestation, 90 percent of which results from clear-cutting for animal agriculture.

Interactions with Wildlife

Before humans, species that lived in forests or their own native habitats existed in relative balance with each other, killing only for the purpose of sustenance – a basic survival need. Humans disrupt this balance – often for "wants." We are seeing rates of extinction comparable to or exceeding those seen during Earth's prior mass extinctions.

Instead of honoring the unique genetic codes that make up the variety of species on Earth, we seem intent on replacing them with just a handful of genetically identical, domesticated plant and animal species that we regularly consume.§§§§§§§§§§§§§§§§§§§§ We replace the natural world with one that is adulterated and tame. We cut lives short (quite literally), and often in inhumane ways.[44,385,]*********************

Now more than ever we must protect wild habitats, mitigate climate change, increase ecological resilience, and improve our own health by changing (and significantly reducing) our consumption habits.

The *Most Effective* Way to Achieve These Goals Is to Reduce Our Intake of Animal Products

While most people are aware that fossil fuels are major sources of greenhouse gases, far fewer people are aware of just how much the foods we eat contribute directly to climate change, sea-level rise, ocean acidification, and global extinctions, all of which were covered in Part 1 of this book.

By replacing animal products – such as meat and dairy – with plant-based alternatives, we can eliminate up to 70 percent of all greenhouse gas emissions from foods and reduce water use by up to 75 percent.[34,386] Dietary change is significantly more effective at reducing greenhouse gas emissions than commonly promoted strategies such as recycling or changing to energy-efficient light bulbs, and it can also reduce, if not stop, future deforestation.[34,364] This swap alone is one of the most meaningful, effective, and proactive things we can do to benefit our health and the environment simultaneously.

---

§§§§§§§§§§§§§§§§§§§§ Refer back to Figure 5.1 for more detail on where genetic diversity is found among animals species.
********************* See the informative graphic at www.four-paws.us/campaigns-topics/topics/farm-animals/life-expectancy.

Furthermore, this swap could save more than 320,000 human lives every year from premature death due to obesity and/or chronic disease. This represents a $20–30 trillion saving worldwide in medical care costs *every year*.[34]

The best part about this is that it is something we can act on right now; we do not need government intervention to make it happen – which is probably a good thing, since in far too many conferences on climate change the topic of dietary impact is frequently ignored or simply not addressed. On the subject of climate change, too often the focus remains on less impactful actions such as changing to energy-efficient light bulbs, recycling, hanging clothes out to dry instead of using a tumble dryer, or switching to a hybrid car.††††††††††††††††††††††† Yet the fact is that simple dietary changes would be far more effective at reducing greenhouse gas emissions (between four and eight times more effective).[364]

Currently, food systems around the world produce enough calories to meet the needs of every person and domesticated animal on Earth.[387,388],‡‡‡‡‡‡‡‡‡‡‡‡‡‡‡‡‡‡‡‡‡‡‡ Developing countries are home to nearly 80 percent of the world's population, 98 percent of the world's hungry, and 78 percent of the world's harvested croplands. As their populations grow, it will be increasingly difficult to keep up with the increased demand for animal foods without also increasing deforestation or significantly harming the remaining wild places on Earth.[389]

Therefore, it is simply unethical to feed billions of domesticated animals the calories and protein (from plants) that could feed several billions more people and protect the environment all at once.[390] The environmental and societal value of switching from animal-based to plant-based foods is enormous.

If we continue on the path we are on, where more people eat a diet that looks like "America's diet" – high in animal products, highly processed, and full of palm oil – we will lose far too many species, damage the land and oceans, and will need four Earths to feed us

---

††††††††††††††††††††††† It is not clear to me whether this is because of "lack of knowledge or lack of resolve" to guide people to a more sustainable diet or whether it is due to effective (and expensive) lobbying by meat/dairy industries.
‡‡‡‡‡‡‡‡‡‡‡‡‡‡‡‡‡‡‡‡‡ This does not mean, however, that foods are well or evenly distributed. There are more than 800 million people around the world who are malnourished, hungry, or food insecure and unable to meet their nutritional needs.

all.[391,392,393,394,395,]§§§§§§§§§§§§§§§§§§§§§§§§ But we do not have four Earths, we only have this one.

Why aren't more people talking about this?

While scientists and technologists are hard at work finding ways to reduce greenhouse gas emissions from fossil fuels, we are not powerless to make a difference, nor must we wait for a solution. We already have one – perhaps the most impactful one. We just need the will and perhaps the guidance to make use of it.

For anyone who believes that switching to a plant-based diet is too hard or too expensive, it is my goal over the next several pages and in other "recipes" in this section of the book to give you a brief primer and template that can guide you as you start or continue on this journey.***********************

As a dietitian who has spent her life educating and working with patients on how to eat more healthfully, I will now pass on to you some of that practical know-how. And for those of you who may prefer a more prescriptive or guided approach, I present below a week's worth of meal-time examples (especially considering the COVID-19 pandemic) that can help get you started on a plant-based diet that is both doable and easy. First I present some general tips, and then I provide some sample menu plans and grocery shopping tips.

This is a choice – one that you can make today – to help your health and our planet.

General Tips

For my general tips, I discuss how to replace the food items that are most detrimental to your health or the environment with healthier alternatives. For example, I generally recommend that you replace dairy products with nondairy alternatives, as you will see in the examples below. There are three reasons for this:

First, health – dairy products contain proteins that can be detrimental to your health. The major protein component of cow's

---

§§§§§§§§§§§§§§§§§§§§§§§§ For a fairly extensive list of products made with palm oil, see www .deforestationeducation.com/products-that-contain-palm-oil.php. If we continue on the path we are on, it is very likely that orangutans could go extinct in our lifetime (Scientific American, 2020).
*********************** There are many ways in which a whole foods, plant-based diet can be less expensive than a standard "American" diet.

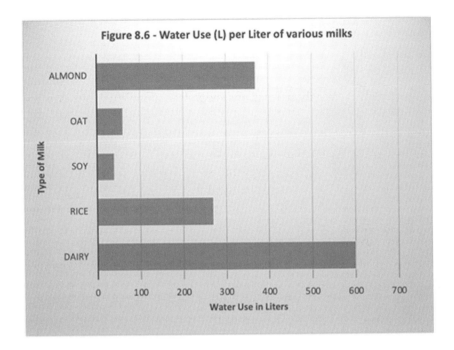

Figure 8.6 Water use per liter of various milks
Data adapted from Poore & Nemecek (2018) and also from Guibourg & Briggs (2019)

milk – casein – has been shown to promote the growth of certain tumors.[396]

Second, also health – dairy products contain naturally occurring hormones such as estrogen, progesterone, and even testosterone. These are the same hormones we produce in our own bodies. Cow's milk contains these hormones because cows are mammals just like we are. Personally, I am just not particularly comfortable with the idea of drinking hormones from another animal and have always given my son nondairy alternatives.[397]

Third, the environment – producing dairy products is extremely resource intensive. One liter of cow's milk requires 600 liters of water to produce, whereas the same liter of soy milk only requires roughly 30–50 liters (5–10 percent) of water to produce (almond milk requires more water than soy milk, but only about 50–60 percent as much as cow's milk; see Figure 8.6).[398],††††††††††††††††††††††††††

---

†††††††††††††††††††††††††† The water used to produce milk includes both the water that cows drink as well as the amount of water that is needed to produce their feed. (I will add that there

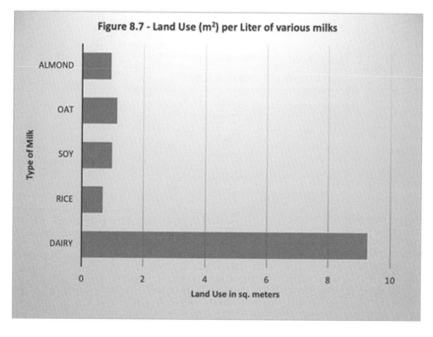

Figure 8.7 Land use per liter of various milks
Data adapted from Poore & Nemecek (2018) and also from Guibourg & Briggs (2019)

The amount of land required to produce various milks is on a similar scale. To produce 1 liter of cow's milk requires nearly 9.3 m² of land. In the course of a year, one cow requires roughly 81,000 m² (20 acres) of land to produce nearly 2,300 gallons (8,700 liters) of milk – which is about the same amount of land as 20 soccer (football) fields would fill. Plant-based milks require an average of less than 10 percent that amount of land (see Figure 8.7).[398]

Finally, dairy produces roughly 3 kg of carbon dioxide equivalents per liter compared to one-quarter to one-third that amount for many nondairy milk alternatives (see Figure 8.8).

Like dairy, there are a number of alternatives to meat products that I recommend for health and environmental reasons. It is well known that the saturated fat in meat products is detrimental to health and associated with inflammation, heart disease, and increased risks of

are numerous welfare issues with both cow's milk – as cows are forcibly impregnated and their calves taken from them immediately after birth – and almond milk – which requires bees for pollination, putting the bees' lives at risk due to pesticide use and/or overworking them.)

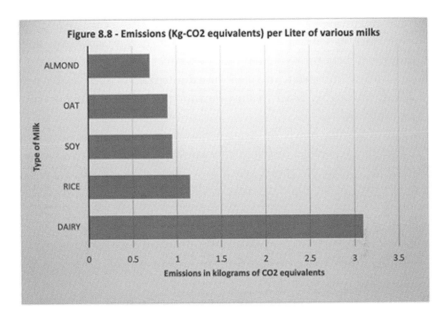

Figure 8.8 Emissions per liter of various milks
Data adapted from Poore & Nemecek (2018) and also from Guibourg & Briggs (2019)

certain types of cancer, including colon cancer.[399] And, as described previously, meat products produce huge amounts of greenhouse gas emissions and require significantly more water to produce compared to plant-based alternatives. If you refer back to Figure 8.3, you can see that dairy and beef products are some of the most resource-intensive (and environmentally destructive) foods.[400]

Keeping these examples in mind, here are some tips you can start with:

Instead of using cow's milk, use plant-based nondairy milk alternatives such as soy/soya milk, oat milk, coconut milk, pea milk, or almond milk (there are others as well). You can even make your own.

Instead of dairy yogurt, use a nondairy alternative such as soy milk yogurt, oat milk yogurt, cashew milk yogurt, almond milk yogurt, or coconut milk yogurt.

Instead of butter on toast, which is full of saturated fats and is water-intensive, use Marmite, vegemite, hummus, avocado, or nut/seed butters.

Instead of a meat burger, try a plant-based alternative, including Beyond Meat, Impossible Meat, or another similar brand.

Instead of a meat chili, try a vegetarian chili that uses beans and hearty vegetables or extra-firm tofu, which is still deeply satisfying, delicious, and better for you and the environment.

Instead of ground meat, try plant-based alternatives such as Beyond Meat, Impossible Meat, beans, seitan, tofu, or other similar types or brands.

Instead of tortilla bowls or Asian bowls with meat, try grilled tofu, edamame, vegetable, or seitan alternatives.

Instead of meat or cheese enchiladas or tacos, try bean, tofu, alternative meat, nondairy cheese alternatives, or grilled vegetable fillings.

Instead of eggs (which require sixty gallons of water *each* to produce), try tofu, silkened tofu, or other alternative "egg substitute" products, including flax meal. If you feel like scrambled eggs, you can try scrambled tofu instead! It's a dead ringer!

Instead of raw fish sushi, try vegetable or avocado sushi.

With these substitutions you can still enjoy many of the same flavors, cultural foods, and experiences, but with a healthier and more environmentally friendly spin or taste. What's more, you will feel really good about it because you know that what you are eating is better for your health and reduces your risk for chronic disease. These are really good things!

I do realize that some of you may have concerns about whether or not you can get enough protein with a plant-based diet – and this is because I am frequently asked this question. The answer is yes, you absolutely can get more than enough protein (and other nutrients) with a plant-based diet. And, for most of us, we do not need nearly as much protein as we think we do!

As a quick example, a generally healthy 140-pound (64-kg) female only requires 50–65 grams of protein per day, while a generally healthy 180-pound (81-kg) male only requires 65–80 grams of protein per day.‡‡‡‡‡‡‡‡‡‡‡‡‡‡‡‡‡‡‡‡‡‡‡‡

Given these relatively low protein requirements (compared to the incorrect assumptions about how much protein we need), it is quite

---

‡‡‡‡‡‡‡‡‡‡‡‡‡‡‡‡‡‡‡‡‡‡‡‡ There are some medical conditions that require more or less protein, and *I strongly recommend* that you seek the advice of a physician or registered dietitian to guide you on how much protein you need based on your own personal medical history.

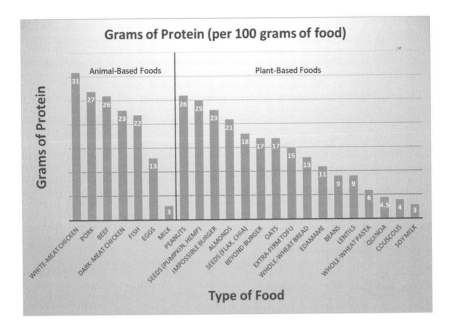

Figure 8.9 Protein contents of various types of foods

simple to obtain these amounts from plant-based foods. Please see Figure 8.9 to gain a better understanding of how much protein is in several different types of foods, both animal-based and plant-based.

For additional and more specific examples of how to successfully follow a plant-based diet, here are some tips!

Breakfast Tips

- Whole-grain cereal with soy milk, almond milk, or other milk alternatives, and nuts.
- Whole-wheat pancakes can be made without the use of eggs by substituting silkened tofu, and they can be prepared with alternative milks.
- Whole-wheat toast or bagels with hummus or peanut butter.
- Oatmeal prepared with soy milk, almond milk, or water and served with flaxseed or nuts and berries.
- Scrambled tofu with sautéed vegetables.
- Nondairy yogurt with almonds, berries, cashews, and whole-grain granola.
- Whole-grain toast with avocado, roasted bell pepper, and pureed edamame.

Lunch Tips
- Vegetarian chili with whole-grain bread.
- Grilled-vegetable panini with edamame hummus on whole-grain bread.
- Vegetable barley soup with whole-wheat pita bread.
- Veggie burger on a whole-wheat bun with onion, lettuce, tomato, mustard, and ketchup.
- Bibimbap (a Korean dish) with tofu and black or brown rice.
- Black bean and corn tortilla enchiladas.
- Vegetarian tacos.

Dinner Tips
- Vegetarian tofu fajitas with corn tortillas.
- Vegetarian pizza (you can make your own whole-wheat pizza crust if you like, or use cauliflower pizza crusts or other gluten-free pizza crusts) topped with a rainbow of vegetables. You can even buy vegan cheese for a topping if you like.
- Whole-wheat pasta with tomato sauce packed with vegetables such as zucchini, mushrooms, onions, and garlic.
- Split pea or lentil soup with whole-wheat French bread or gluten-free bread.
- Vegetarian lasagna (tomato-based) made with whole-grain or gluten-free noodles.
- Quinoa salad with ratatouille.
- Roasted bell peppers stuffed with a combination of pearled barley, kale, and sautéed onions and mushrooms.

Now that I have given you some general tips, see Tables 8.1 and 8.2 for a couple of "meal plans" to guide you through a week's-worth of meals. We use meal plans in our home as they help us to know exactly how much to buy, especially in the middle of a pandemic when we don't want to waste food or spend longer than necessary shopping for food.

As an alternative to cooking new items every day of the week, you can do what we do in my household, which is to cook extra-large portions of more time-intensive meals on the weekends when we have the time – and, frankly, the energy – to cook, and then reserve the rest for later on in the week. I find that reheated leftovers (especially soups and stews) often taste even better because the spices and flavors have had the chance to marry together.

Table 8.1 Example seven-day meal and grocery list plan (for a standard North American or UK/Australian diet)

	Saturday	Sunday	Monday	Tuesday	Wednesday	Thursday	Friday
Breakfast	Whole-grain cereal w/ nondairy alternative milk and nuts	Whole-grain pancakes (American or British style) w/ syrup or fruit on top	Whole-grain toast w/ hummus, peanut butter, Marmite, or jam	Oatmeal or porridge topped w/ flax, nuts, berries, or jam	Scrambled tofu w/ sautéed vegetables	Nondairy yogurt w/ nuts, berries, or granola	Whole-grain toast w/ bell pepper, avocado, or edamame
Lunch	Vegetarian chili or soup w/ whole-grain bread	Grilled-vegetable and edamame hummus panini	Vegetable-barley soup w/ whole-wheat pita bread	Veggie burger on a whole-grain bun served w/ side of veggies and/or chips	Bibimbap w/ tofu and black or brown rice	Black bean, tofu, and corn tortilla enchiladas	Grilled vegetable tacos w/ a side of salsa or guacamole
Dinner	Vegetable soup and salad	Vegetarian pizza w/ fruit smoothie for dessert	Whole-wheat pasta w/ tomato primavera sauce	Split pea or lentil soup w/ whole-grain bread	Vegetarian lasagna w/ whole-grain noodles	Quinoa salad w/ ratatouille	Stuffed/roasted bell peppers
Suggested grocery list							
Whole-grain cereal		Whole-wheat flour		Barley[M,F]		Split peas/lentils	
Alternative milk[Sa,T]		Edamame[Su,F]		Whole-wheat pita bread		Whole-grain lasagna noodles	
Veggie chili		Zucchini[Su,W,R]		Whole-wheat pasta		Nondairy yogurt	

Table 8.1 cont'd

Whole-grain bread[Su,M,F]	Mushrooms[Su,M,F]	Granola
Tofu[Sa,W,F]	Silken tofu	Black beans
Onions[Sa,Su,T,R,F]	Vegan cheese	Quinoa
Bell peppers[Sa,Su,F]	Hummus[Su,M]	Eggplant (for ratatouille)
Corn tortillas[Sa,R,F]	Peanut butter	Avocado
Marmite	Jam	Corn tortilla

Several items may be used on several days of the week (annotated with superscript M, T, W, R, F, Sa and/or Su), but will only be listed once in the chart. Nonperishable goods such as cereals, raw split peas, peanut butter, oats, and flax can be used for multiple weeks.

Table 8.2 Example seven-day meal and grocery list plan (for a typical European/Italian-style diet)

	Saturday	Sunday	Monday	Tuesday	Wednesday	Thursday	Friday
Breakfast	Whole-grain crepe or English-style pancake w/ fruit	Hash browns/fried potatoes served w/ whole-grain toast, baked beans, fried tomatoes, and mushrooms	Muesli w/ nondairy milk, fruit, and nuts	Nondairy yogurt w/ granola on top	Rustic loaf w/ jam, peanut butter, or plant-based cream cheese	Baguette w/ jam or Marmite	Cereal w/ nondairy milk, fruit, and nuts
Lunch	Potato pierogi (Polish dish) w/ a side of fruit	Vegetable pizza (Italian)	Greek salad w/ a plant-based (vegan) cheese	Vegan halloumi and watermelon (Cypriot specialty)	Vegetarian caldo verde (green soup – Portuguese dish)	Bami goreng (Indonesian specialty served in The Netherlands)	Bucatini arrabiata (Italian pasta in a spicy tomato sauce)
Dinner	Borscht w/ a side salad	Spaetzle w/ roasted vegetables	Tofu caprese salad	Greek yahni (potato stew)	Plant-based meat-stuffed cabbage	Tofu and spinach borek	Falafel w/ pita bread
Suggested grocery list							
	Whole-grain flour	Nondairy milk		Fruit (fresh/frozen)		Potatoes	Whole-grain toast
	Baked beans	Tomatoes		Mushrooms		Nondairy yogurt	Granola
	Whole-grain baguette	Jam/peanut butter		Vegan cream cheese		Nuts	Marmite

Table 8.2 cont'd

Whole-grain cereal	Pierogi (can also make)	Vegetable pizza	Vegan feta/ricotta	Vegan halloumi
Watermelon	Caldo verde (can also make)	Bami goreng (can also make)	Bucatini/pasta	Arrabiata sauce
Borscht (can also make)	Spaetzle (can also make)	Onions (to roast)	Zucchini (to roast)	Mushrooms (to roast)
Tomatoes (for roasting)	Tofu (for various salads)	Basil (for caprese salad)	Cabbage	Plant-based meats
Spinach	Filo dough	Falafel (can also make)	Pita bread	Vegan ricotta cheese

Thus, Table 8.3 provides an example of what *our* weekly meal planning grid looks like, which both cuts down on the total amount of cooking that has to be done in a given week and also uses up the most perishable food items (in cooking) at the beginning of the week when they are still freshest (thus reducing food waste).

Now that you see how meal planning can help make this idea easier and more doable, I recommend looking up vegetarian and vegan recipes online (or in a cookbook) that are filling, make several servings, require only a handful of easy-to-acquire ingredients, and, of course, contain flavors that you enjoy. My examples are just that – examples. Feel free to insert your own! If you find any really good ones, keep them on standby and refer back to them on a regular basis as we do in our own household.

With your new meal planning ideas and recipes in mind, also be on the lookout for organic or fair trade food and beverage items (within your budget) wherever and whenever possible. The reasons to do so are outlined in Recipe 2.

Recipe 2 Buy Organic or Fair Trade

When deciding which foods or beverages to purchase and consume, there are many variables to consider aside from just "taste." Some of these variables include whether the items were grown using regenerative, organic, or conventional methods, whether or not the items were sustainably sourced, and whether or not the items are certified fair trade.

Buying foods and beverages is no longer just a decision about whether you feel like having an apple, a banana, or an orange, nor is buying coffee just about whether it is caffeinated or decaffeinated. These days, the variety of choices available in supermarkets and the reasons behind why we make certain choices can feel mind-boggling.

As a quick primer, the following are a few definitions to be aware of for *your* next shopping trip:

Ethical shopping and consumerism: The idea surrounding "ethical shopping" involves several characteristics, including:[401]

(1) "Positive buying," which is making the decision to purchase items because they are from a fair trade, organic, or cruelty-free source. These designations and/or certifications are *the* most important

Table 8.3 What our weekly meal plan actually looks like

	Saturday	Sunday	Monday	Tuesday	Wednesday	Thursday	Friday
Breakfast	Vegan whole-grain pancakes w/ maple syrup (make extras)	Vegan whole-grain waffles or cinnamon buns (make extra)	Whole-grain cereal w/ soy or almond milk and fruit	Leftover pancakes	Whole-grain cereal w/ soy milk or almond milk and fruit	Leftover waffles or cinnamon buns	Whole-grain cereal w/ soy milk or almond milk and fruit
Lunch	Grilled vegetable sandwich on whole-grain bread	Hummus and whole-grain pita bread	Peanut butter or almond butter and banana sandwich	Grilled vegetable sandwich on whole-grain bread	Hummus and whole-grain pita bread	Peanut butter or almond butter and banana sandwich	Takeout lunch (e.g. vegetable tacos from your favorite Mexican restaurant)
Dinner	Vegan enchiladas (we make enough for 3–4 nights)	Vegetable and whole-grain pasta soup (we make 12 quarts or more)	Vegan burgers (e.g. Impossible Burger or Beyond Burger) on whole-grain buns	Leftover enchiladas	Leftover soup	Vegan burgers or leftover enchiladas again	Leftover soup

considerations for some ethical shoppers, as they directly support progressive companies that actively engage in these practices.

(2) "Negative purchasing," which, just as it sounds, is the act of *purposefully avoiding* buying products that you disapprove of because of how the product is produced or how it is "raised." For example, some consumers will not buy eggs that have been laid by battery cage-raised hens.§§§§§§§§§§§§§§§§§§§§§§§§§§

(3) "Company-based purchasing" – this type of shopping targets businesses as a whole. For example, you may choose to boycott buying anything from a large multinational company and all of its subsidiaries as a way to compel the company to change its ways of doing business (this may be especially the case when the business works with small farmers in developing countries). Typically, this method only works when large numbers of people (e.g. an entire city or an entire country) do this. But it starts with you having awareness and being a more educated consumer.

Conventional agriculture farming methods: Conventional farming generally is associated with modern, high-input agriculture. This often includes the use of synthetic chemical fertilizers, fungicides, insecticides, and herbicides; it may also represent the predominant agricultural practices used in a particular region.[402] Conventional farming methods also tend to use large volumes of water since tilling occurs on a regular (often annual) basis. Tilling does the opposite of regenerative agriculture and causes loss of topsoil, water runoff, and desertification of the surrounding farm area (for more on this, see Recipe 6).[367]

Organic agriculture farming methods: Organic farming prohibits the use of synthetic fertilizers and pesticides, requires certification bodies to certify producers based on a set of production standards, and tends to rely more on healthy living systems, taking advantage of biological diversity and the recycling of nutrients.[402] However, this is not always the case, as there are some organic farms that till on a regular basis.

---

§§§§§§§§§§§§§§§§§§§§§§§§§ Battery cage-raised hens often are crowded three to eight hens in a cage. Each hen is given a space no larger than 550 cm² (85.2 in²; roughly the size of an A4 sheet of paper) to live out her whole life (https://safe.org.nz/our-work/animals-in-need/hens-2/standard-battery-cages).

Worth noting is that the act of tilling, whether on a conventional farming system or an organic farming system, reduces organic carbon levels (and moisture) in soil (as compared to regenerative, no-till farming methods). Tilling predisposes soil to emit carbon dioxide into the atmosphere, lose moisture, and hold on to fewer nutrients, and it may lead to negative environmental consequences.[367,403]

Sustainable and regenerative agriculture farming methods: Agricultural farming methods are considered to be sustainable when current and future food demands can be met without unnecessarily compromising economic, ecological, or sociopolitical needs.********************* Sustainable farming practices offer three major benefits: economic stability, environmental sustainability, and social sustainability. Sustainable agriculture practices – and, even more effectively, regenerative farming practices – have the ability to increase food production without increasing environmental degradation. This is achieved by improving the physical structure of soil, reducing soil erosion, and improving the soil's water storage capacity and nutrient levels through organic and regenerative farming techniques and conservation measures, which include the use of composted and organic compounds that enrich the soil and do not deplete it.[404]

Locally grown: Local food markets typically involve smaller farms that grow or raise a variety of products and have short supply chains. Farmers often perform all of their own marketing functions, including the storage, packaging, transportation, distribution, and advertising of their products. Local food systems tend to enhance employment and income within their own communities as well as reduce the amount of energy required and greenhouse gases emitted to transport products from farm to market.[405]

Fair trade: The basic tenet of fair trade is that it supports responsible companies, empowers farmers, workers, and fishers, and protects the environment by paying workers a fair price for their products and the work they do, regardless of any volatility in market prices.

********************* The sociopolitical dimensions of sustainable agriculture include: food security, hunger, nutrition, food contamination, and obesity (human health); participation, social capital, community cohesion, and governance (democracy); environmental and food justice (equity, justice, and ethics); the capacity of a system to absorb disturbances and reorganize while undergoing change so as to retain essentially the same function (resiliency and vulnerability); and cultural practices, languages spoken, and indigenous and hybrid ecological knowledge systems (biological and cultural diversity); see Bacon et al. (2012).

Fair trade producers tend to engage in environmental stewardship to help keep the planet healthy by prohibiting the use of harmful chemicals and by taking measures to protect and regenerate natural resources.[406]

When done correctly, fair trade provides sufficient income, health, and safety to farmers, workers, and fishers who often are the people who are most vulnerable to fluctuations in global markets as the price paid to farmers is frequently driven down by large companies and/or their subsidiaries, while the companies themselves make extreme profits.

In a fair trade system, cooperatives of farmers, workers, or fishers work together to create a sustainable, often organic, and healthy product using transparent supply chains in exchange for fair wages and additional money that go into helping to meet their communities' social, economic, and environmental needs.[406]

Although there are a number of definitions, certifications, and packaging symbols to choose from, as described above, and it can feel confusing and overwhelming, it is only when we have a better awareness of what the various symbols on food packaging mean that we can truly make better-informed choices with our buying power.††††††††††††††††††††††††††

While many of you have likely heard about the health benefits of organic products – from fruits and vegetables to grains and legumes – I'm sure there are also many of you who may be less familiar with the environmental advantages that are associated with regenerative, organic, fair trade, or sustainable farming.

Some of the environmental advantages associated with organic – especially regenerative – farming include: increased biological diversity both on the farm and surrounding the farm; increased use of crop rotation methods, which have many benefits for soil health and quality; prohibited use of chemical herbicides and insecticides, which benefits both public health and the environment; and more mindful management of noncropped margin habitats, which can be beneficial for both animal and plant species that are not associated with the farming process.[407]

Sadly, and incorrectly, organic agriculture is often criticized for producing lower yields than conventional agriculture. But yield differences actually are highly dependent on the product being grown, type of

†††††††††††††††††††††††††† For some examples of what seals and symbols on food packaging might look like and what they mean, please see www.farmaid.org/food-labels-explained.

system used (tilling versus regenerative), soil quality, and site characteristics. Therefore, when good management practices and certain crop varieties are used, organic systems can very nearly match, if not exceed, conventional systems in how much they yield.[408]

Regenerative farming, which is frequently organic (or nearly organic), is a system of farming aimed at producing food while causing the least amount of harm to ecosystems, animals, or humans.[367] Regenerative farming is proposed as a solution to the environmental ills of conventional farming by reducing, if not eliminating entirely, the use of phosphorous- and nitrogen-rich fertilizers and other toxic chemicals, which are extremely harmful to the environment and to health when they accumulate in the body.[409]

Regenerative and organic farming techniques are more sustainable over the long term and lend themselves nicely to positive and ethical consumerism.

Although the price of organic products still is higher than the price of conventional products throughout much of the world, the gap is closing.[410] Premiums on organic products tend to help small farmers in developing countries because the prices for organic products are more stable in the short run, implying less risk and more certainty. This is also the case for many fair trade products.[411,412] It is also important to remember that the environmental cost – in terms of land degradation, climate change, and water use, among others – is not added into the cost of foods produced from conventional farming systems, thereby making them less expensive (to the consumer) than they ought to be.

Ultimately, *you* decide which foods and beverages you buy. The next time you go to the grocery store, examine some food packages, fresh produce, and their labels. Look to see whether any of these products have fair trade, regenerative farming, organic, or other certified designations or symbols on them. Look at where they were manufactured or packaged. Also look at the price, as some fair trade or organic items may cost no more than the same conventional item. If you are so inclined, look up the company on a third-party website that rates the ethics or sustainability level of that company, because you never know what you may find or what you may learn.[‡‡‡‡‡‡‡‡‡‡‡‡‡‡‡‡‡‡‡‡‡‡‡‡‡] I think it is worth making the extra effort

---

[‡‡‡‡‡‡‡‡‡‡‡‡‡‡‡‡‡‡‡‡‡‡‡‡‡] You can try websites such as www.ethical.org.au or www.guide.ethical .org.au and see what you find!

for better-tasting, healthier, and environmentally friendly foods, and so does my son.

With healthy and sustainable foods in mind, read on about why it is important to get your omega-3 fats, but also why you can ditch the fish!

Recipe 3 Eat More Omegas (but Ditch the Fish)!

As discussed in Part 1, it is difficult to imagine how current levels of overfishing can continue, or be sustainable business-as-usual.

Unfortunately, there are too many advertisements touting the supposed health benefits of omega-3 fatty acid supplements – especially those that come from fish oil. Additionally, too many medical doctors, dietitians, naturopaths, osteopaths, and the supplement industry itself will leave you believing that we need more, more, and more of these omega-3 fats in our diet to prevent heart disease, stroke, diabetes, or cancer, and that the only way to get them is by eating more fish or taking more fish oil supplements.

This is simply not true.

While we definitely require omega-3 fatty acids from our diet for optimal health, as our bodies cannot make them on their own, we do *not* need to get them from fish or fish oils (which are often very expensive).

Quick Background on Omega-3 Fatty Acids

Omega-3 fatty acids are known as *essential* fatty acids, specifically because we cannot make them on our own and because we *must* obtain them from the foods that we eat. And while we require omega-3 fatty acids from our diet, there is no requirement to get them from fish or fish oil.

Instead, we can obtain more than a sufficient amount of omega-3 fatty acids from plant-based sources, including algae, walnuts, flaxseeds, chia seeds, and hemp, to name a few. Moreover, plant-based sources of omega-3 fatty acids generally are safer than those from fish because they do not have the same opportunities to bioaccumulate polychlorinated biphenyls, mercury, or other plastic by-products and petrochemicals that fish have.

In fact, research supports a whole foods, plant-based diet, as described in Recipe 1, as a better, cheaper, and more effective method to prevent and mitigate chronic diseases than taking high doses of fish oils.[26,27]

In one meta-analysis of eighteen randomized controlled clinical trials§§§§§§§§§§§§§§§§§§§§§§§§§§§§ that looked at the use of fish oils as a preventive measure, only two of the studies demonstrated *any* positive health benefits from taking fish oils. The other sixteen studies did not demonstrate any benefits or proof of efficacy.[413] This analysis looked specifically at the use of fish oils for cardiovascular, neurocognitive, ophthalmic, and inflammatory disorders. What's more, the same meta-analysis found that the vast majority of people who take fish oil supplements are doing so without explicit instructions from their medical provider. Basically, the meta-analysis concluded that most people who take fish oil supplements are doing so because they *believe* they will be beneficial for their health. There is simply too little proof of that.

In another study, the American Heart Association found that patients taking a daily four-gram dose of omega-3 fatty acids from fish oil fared better after a heart attack than those taking a placebo. However, it is important to note that this study addressed the fact that taking four grams of fish oil per day is a very hard pill to stomach, even for the hardiest of us.[414] In order to take in four grams of omega-3 fatty acids from fish oil, you would need to consume anywhere between six to eight large fish oil pills *every day* – and possibly more. That's because a typical fish oil supplement or pill contains between 400 and 1,200 mg of fish oil, but only one-third to one-half of each fish oil pill contains the all-important omega-3 fatty acids that we hear about most often: eicosapentaenoic acid (EPA) and docosahexaenoic acid (DHA).************************

Unfortunately, consuming that much fish oil can lead to a number of untoward and uncomfortable side effects, including belching,

---

§§§§§§§§§§§§§§§§§§§§§§§§ Randomized controlled clinical trials are the gold standard of medical studies in which the study design randomly assigns participants to one of two groups: an experimental group or a control group. As the study is conducted, the only expected difference between the two groups is the outcome that is being studied.
************************ Take a look at fish oil pill bottles the next time you are at the store to see how much EPA and DHA is *actually* in each pill. While there *are* some superpotent fish oil pills available on the market, they are relatively few and far between. Most require megadoses.

bad breath, heartburn, nausea, loose stools, and nosebleeds, to name a few, making this type of regimen extremely difficult to comply with.[415]

Cardiologist Ilan Kedan – Division Chief of Cardiology at Cedars Sinai Medical Center in Los Angeles – does not recommend the use of fish oil to his patients.[†††††††††††††††††††††††††††] He states, "There is no convincing data to suggest effectiveness at lower doses, and most patients are not going to comply with or remember to take that many fish-oil pills each day, as they are already taking several pills." Many individuals who have heart disease also have other chronic diseases, and "pill fatigue" is a real thing – I see it all the time with many of my patients who are in hospital long term.

Another potential concern with eating fish and taking fish oil supplements is what else we might be taking in when we consume them. As discussed in depth in Chapter 4, there is a lot of plastic debris in the oceans, along with mercury and other industrial and environmental chemical wastes.[415] Fish and other marine creatures consume and take in a lot of these tiny plastic particles or other toxic chemicals that are embedded in plastics such as polychlorinated biphenyls, bisphenol A (BPA), and dichlorodiphenyltrichloroethane (DDT), among others. These chemicals get stored and accumulate in the fish's fats, and eventually end up in our bodies when we consume them.

One study found that some virgin (unrefined) fish oil supplements may contain some of these toxic chemicals.[416] Do you really want them in *your* body?!

Examples of Plant-Based Omega-3 Fatty Acids

Fortunately, there are other options available for us to obtain the essential omega-3 fatty acids that we need from our diet and that are healthier alternatives to fish oil. One such example is algae.

Algae form the foundation of the marine food web. Algae provide needed oxygen to the oceans and the atmosphere, and they also are *the primary producers* of omega-3 fatty acids, which fish eat. Without algae or other forms of microbial production of omega-3 fatty acids, fish would not be the "source" of omega-3 fatty acids that they are.

---

†††††††††††††††††††††††††††† https://bio.csmc.edu/view/3670/Ilan-Kedan.aspx

Algae represent a more sustainable alternative to fish oils. Certain algae species produce EPA and DHA long-chain omega-3 fatty acids and already are used in certain nutritional products, including infant formulas.[417,418,419,420,421] If they are good enough for your baby, they are good enough for you.

Additional sources of plant-based omega-3 fatty acids include walnuts and walnut oil, flaxseeds and flaxseed oil, hemp seeds and hempseed oil, and chia seeds and chia-seed oil, among others. The plant-based omega-3 fatty acids found in these sources come from alpha-linolenic acid (ALA), which our bodies convert into EPA and DHA.[422,423] ALA is safe, viable, more sustainable, and also free of the plastic and chemical contamination risks associated with fish and fish oils.

Dr. Marion Nestle – Paulette Goddard Professor of Nutrition, Food Studies, and Public Health at New York University – a frequently cited expert on food and nutrition, told me that "Most of the evidence for the benefits of omega-3s come from studies of fish consumption," such as those looking at the Mediterranean-style diet where fairly high levels of fatty fish are regularly consumed. Dr. Nestle went on to say, "The evidence from fish-oil supplement studies [however] is decidedly mixed, and when it comes to fish oil, current intake levels may not be terribly effective."‡‡‡‡‡‡‡‡‡‡‡‡‡‡‡‡‡‡‡‡‡‡‡‡‡

What is perhaps most interesting when it comes to omega-3 fats is that ALA is the only one that has an established recommended intake level; EPA and DHA do not. That is because ALA truly *is* the only omega-3 fatty acid that is essential. We can synthesize EPA and DHA within our own bodies from ALA.[424] The amount of ALA we need each day (the adequate intake level) is 1.1–1.6 grams/day for adults and 0.5–1.6 grams/day for children from birth to 18 years of age (see Table 8.4).

These intake levels can easily be obtained from one to two tablespoons of any of the nut and seed oils listed earlier, as well as from soybean oil, pumpkin seeds, and canola oil.[425]

However, if you are looking for an easier, more efficient choice (that does not require your body to transform ALA into EPA or DHA), you *can* buy plant-based sources of DHA. As mentioned earlier, marine

---

‡‡‡‡‡‡‡‡‡‡‡‡‡‡‡‡‡‡‡‡‡‡‡‡‡ I had the pleasure of interviewing Dr. Nestle for a piece I wrote on this topic for *Self* magazine several years back.

Table 8.4 Adequate intakes for omega-3 fatty acids

Age	Male	Female	Pregnancy	Lactation
Birth to 6 months[a]	0.5 g	0.5 g		
7–12 months[a]	0.5 g	0.5 g		
1–3 years[b]	0.7 g	0.7 g		
4–8 years[b]	0.9 g	0.9 g		
9–13 years[b]	1.2 g	1.0 g		
14–18 years[b]	1.6 g	1.1 g	1.4 g	1.3 g
19-50 years[b]	1.6 g	1.1 g	1.4 g	1.3 g
51+ years[b]	1.6 g	1.1 g		

Adapted from the US National Institutes of Health Office of Dietary Supplements.[424]
[a]As total omega-3 fatty acids.
[b]As alpha-linolenic acid.

algae and algal oils are wonderful sources of plant-based DHA, and they actually are where fish obtain their DHA in the first place – when they eat marine algae.

Like some of the other plant-based sources of omega-3 fats, a benefit of algal DHA is that it does not carry the same risk of contamination from plastics and other chemicals as fish oil, since algae are at the very bottom of the food chain. Moreover, algal oils do not seem to exhibit the same negative side effects that fish oils do (namely the gastrointestinal side effects), and they have been used in food products for years.[417]

In summary, it is true that we need omega-3 fats for good health, and we most definitely need ALA. However, in the last several years, fish oil's efficacy and, frankly, safety have come into question. As a dietitian who specializes in cardiology, a large proportion of my patients have heart disease, and several are waiting to receive heart transplants or have already received one. While it used to be common, very rarely now do I see patients receiving fish oil supplements. It seems that it is not the "panacea" it was once believed to be; and in fact, the "good fats" that you need for good health, like so much else in nutrition, can more easily and more healthfully be obtained by "eating your veggies."

However, if you do happen to be someone whose doctor has prescribed them fish oil for therapeutic reasons, consider presenting and discussing some of this with your doctor as a potential alternative to fish oils. If, after that discussion, your doctor continues to recommend fish oil for therapeutic purposes, consider seeking out fish oils that are both sustainable *and* safe to consume.

One example of a more sustainable option is fish oils that come from farm-raised fish in marine, offshore environments that use an integrated multitrophic aquaculture system.[426],§§§§§§§§§§§§§§§§§§§§§§§§§§§§§§ Self-contained, multitrophic systems are far more sustainable than traditional aquaculture that uses wild-caught fish feed for a few reasons. First, they eliminate the risk of by-catch of unintended species. Second, they recycle nutrients among the various species within the system. Third, because they are self-contained, they do not pollute the oceans or other natural marine systems. And fourth, they are extremely efficient in their use of resources and reduce or even eliminate the need for wild-caught fish feed (something that is far too common in aquaculture – for more details, see Chapter 3).[426]

While it would be more environmentally beneficial to use algae, plant-based omega-3 oils, and/or offshore multitrophic systems, there are a myriad of potential forces preventing their more extensive use. Some of these forces include the costs of production and scaling up, inadequate communication regarding the safety and efficacy of these products, and perhaps most importantly the battle against the many strong fisheries lobbies that do not want these alternatives to succeed.[427]

In addition to the uphill battle in disseminating the importance of plant-based sources of omega-3 fats, there are other complicating factors as well. These include the growing popularity of sushi and poke restaurants, which often serve deep-water, high-trophic, and possibly endangered fish species, and, of course, the deeply held belief that fish and fish oils are the best or only sources of healthy omega-3 fats out there.[428]

To combat some of these challenges, fishing companies can look to diversify their portfolios by investing in alternatives, where there are

---

§§§§§§§§§§§§§§§§§§§§§§§§§§§§ An integrated multitrophic aquaculture system is one that mimics a natural food chain found in nature, where the waste of one species becomes the food/nutrients for another species. For an example, see www.dfo-mpo.gc.ca/aquaculture/sci-res/imta-amti/index-eng.htm.

many opportunities for growth, sustainability, and research and development. As an example of this, fishing companies can look at what Danone did to diversify itself.

Danone, a dairy- and yogurt-producing company, purchased WhiteWave, a nondairy milk alternative-producing company, in April 2017. This has been a very successful and profitable investment for Danone as more people turn to nondairy milks.

Similarly, if fishing and fish supplement companies work together with plant-based omega-3 sourcing companies to develop the technology involved in growing sustainable alternatives, they too can benefit from the win–win scenario that Danone has experienced: sustainable omega-3 fatty acid sources, sustainable fisheries, and long-term growth for both.[429] Changes do not need to be at odds with consumer health, the environment, or a company's bottom line; every sector can benefit from these changes.

As more of us switch to healthier and more environmentally friendly products, more companies will see the need to develop and produce these products, which will help drive down their cost. It is you – the consumer – who drives positive change.

To conclude, it is true that we all need omega-3 fatty acids in our diet for optimal health. But it is also true that we do not need to get them from fish or fish oils. We can get healthy omega-3 fatty acids from sustainable, plant-based, or algae-based sources. We do not need megadoses (from fish). *This* is something we can all start doing today, and by making these changes, we can help prevent wasteful by-catch and, perhaps, even food waste.

Recipe 4 Decrease the Amount of Food You Waste

As described in more detail in Chapter 1, nearly 40 percent of all food produced in the United States and nearly one-third of all food produced globally gets discarded every year. Once discarded, food waste undergoes anaerobic processes************************** and releases 4.4 gigatons of carbon into the atmosphere as methane gas.[430]

************************ Anaerobic processes break down biodegradable material in the absence of oxygen.

Putting this into context, the by-products of food waste release enough carbon to be the third largest producer of greenhouse gases in the world after the United States and China.[37]

Food waste occurs all along the supply chain, from the farm to the kitchen. In a typical year, food waste will vary by individual and by country. In Latin America, 11 percent of all foods get discarded, while in North Africa and Central Asia, that figure is 16 percent. Asia discards 31 percent of their foods, Europe as much as 34 percent, and across America, we waste a staggering 39 percent of foods.[431] These figures are tragic, as nearly 11 percent of American adults and 18 percent of American children do not get enough to eat. Sadly, too many others eat unhealthily.[432] In the UK, it is estimated that nearly 6 percent of people aged fifteen or over do not get enough to eat.[433] It is unconscionable to think that so many more people could be fed with the foods that get thrown away.

In low-income countries, food waste occurs primarily during the production process, while in middle- and high-income countries, food waste is driven primarily by retailers and consumers who want their food looking pristine.[431] Disappointingly, in the United States, one-third of harvests are thrown away in packaging houses as a result of not meeting high cosmetic standards – that they be blemish free or perfectly symmetrical – even if they are perfectly safe and healthy to eat.[434]

All of this waste not only produces huge amounts of carbon emissions, but also discards the resources that were used to produce those foods, including 250 km³ of water every year (about as much water as 100 million Olympic swimming pools) and the equivalent of roughly 28 percent of the world's agricultural area.[37,431]

Given that on the entire Earth there is only roughly 100,000 km³ of freshwater for all humans (and animals) to use each year for all activities, including agriculture, industry, hygiene, and drinking, the fact that so much of this valuable resource gets wasted is tragic.[435,]††††††††††††††††††††††††††††† Additionally, intensive farming often leads to desertification,‡‡‡‡‡‡‡‡‡‡‡‡‡‡‡‡‡‡‡‡‡‡‡‡‡ the loss of soil

---

††††††††††††††††††††††††††††† Freshwater makes up only 3 percent of *all* water on Earth.
‡‡‡‡‡‡‡‡‡‡‡‡‡‡‡‡‡‡‡‡‡‡‡‡‡ Desertification is the persistent degradation of dryland ecosystems by human activities – including unsustainable farming, overgrazing, and clear-cutting of land – and by climate change (UN, 2019). Desertification can be reduced or prevented through improved farming methods and soil management, as described in Recipe 2.

nutrients, and the loss of soil itself, making it increasingly difficult to grow the same amount of foods (or even more) on the same land.[367,436]

Perhaps even worse, over 6 billion pounds of fresh produce are left to rot in fields every year because too much is planted or it is damaged due to weather, pests, or disease. The other reason this happens is because market prices may be less than the cost of transportation and labor, making it too costly to move the produce off of farms.[437] As food becomes increasingly difficult to grow as a result of climate change or weather extremes, it will be ever more imperative that we not continue such wasteful behaviors.

Sadly, amid the COVID-19 pandemic, billions of *additional* pounds of food, gallons of milk, and dozens of eggs more than usual were discarded or plowed over in fields. There were several reasons for this, including fewer institutional buyers (e.g. restaurants had to shut down), a change in the need for harvesting and moving crops or farm animals (e.g. processing plants slowed down business due to COVID-19 infections), and a lack of funds to pay farm workers to harvest and move the crops. All of these factors limited the ability to get products to market.[88,89] These supply-chain disruptions were particularly bad in the United States.

Europe saw significant strains on their food supply chains as well, which resulted from logistical hurdles involving farmers and their agricultural inputs, processing plants, shipping, retailers, and others along the supply chain.[438] Africa, parts of Asia, and parts of Latin America also experienced disruptions in their food supply chains during the pandemic, though their problems were more "upstream," in that obtaining appropriate inputs to grow foods became far more expensive or more difficult to source.

No matter how you slice it, all of these issues led to food shortages and rising food prices.[439] It is tragic that the foods that went to waste were unable to get to those who had and still have a great need for it, whether in developed or developing countries, where hunger and food insecurity are prevalent, even in normal times (let alone during a pandemic). Even many months after the first round of lockdowns ended, many *more* people around the world are food insecure and at risk of going hungry. Food banks struggle to keep up, and there is still to this day far too much perfectly edible food going to waste.

In addition to agricultural food waste, as addressed in Chapter 1, industrial fishing methods also produce significant but hidden-from-view waste.

When we go to the grocery store, we have no idea how many additional sea creatures were caught – by accident – and died for each fish or seafood item we purchase. However, it is well documented that industrial fishing boats, nets, and lines frequently capture and kill a wide variety of unintended species, known as by-catch or "discards."[440],§§§§§§§§§§§§§§§§§§§§§§§§§§§§§§ We had a more in-depth look at the lack of sustainability from fish and seafood in Chapters 2 and 3.

All food waste, regardless of its source, has ramifications for the global food supply and sustainability. Around 40–50 percent of total food waste and 51–63 percent of seafood waste that occurs in the United States takes place at the consumer level, in our homes, costing an average family of four between $976 and $2,275 per year.[441] In the UK, the average household throws away 3.4 kg (7.5 pounds) of food each week, or roughly £14.92 ($20) worth of food. This turns out to be roughly £730/$950 each year.

As a whole, the UK produces more than 25 million tonnes of greenhouse gas emissions each year from food that gets wasted.[442] Most of this waste occurs because we served ourselves too much, the food sat on the shelf too long, we simply forgot we had it in the house, or the food spoiled due to it being stored improperly.[443]

In Africa, food waste occurs primarily due to inefficient processing and drying, poor storage, and insufficient infrastructure. In sub-Saharan Africa, post-harvest food losses are worth roughly $4 billion per year, or nearly 25 percent of the total crops harvested (and for some crops, such as fruits, vegetables, and root crops, this may reach up to 50 percent). That's enough food to feed at least 48 million people.[444]

A study that looked at perceptions of food waste during the COVID-19 pandemic indicated that people were more cognizant of the foods they were discarding and worked harder to save, store, and eat leftovers.[433] According to this study, these perceptions and food-saving strategies were driven more by the socioeconomic context (job losses and job insecurity) of the COVID-19 lockdowns than by environmental

---

§§§§§§§§§§§§§§§§§§§§§§§§§§§§§§ For additional information on this, I recommend watching the Netflix film *Seaspiracy*, which documents the large amounts of by-catch that occur with industrial fishing.

or health concerns. This makes a lot of sense, as many people were highly concerned about job loss, spending money, food insecurity, and rising prices associated with strained supply chains. However, this study still demonstrates that you are less likely to waste food if you are more uncertain as to where the next supply is coming from, when it is coming in, and how much it will cost you.

With that being said, there are lessons that can be learned from this study that can be applied in an environmentally minded way. There will be future risks to food security that are due to climate change. There will also be disruptions to supply chains associated with climate change (and those that are not associated with climate change – think of the Suez Canal blockage of March 2021) that may once again cause people to be concerned about where their next meal is coming from and how much it may cost them, ultimately leading to less food waste.[431]

One lesson is that by paying greater attention to the foods we buy, minding how much we really need, improving our storage methods, and really maximizing the lifespan of our foods, we can and will throw out less of our food. Strategizing and increasing our awareness can allow us to save money, help the environment, and improve our own personal food security.

While it is true that we may not be able to affect the supply-chain side of food waste, we absolutely can affect our own personal food waste. Right now, more than ever, when we are concerned about our wallets, our livelihoods, and our health and well-being, it is important that we do good for ourselves (and also help the environment simultaneously).[445]

Here are a few ways in which we can all reduce and perhaps even stop our own food waste.

Plan Meals Ahead, Create a Grocery List, and Stick to It!

Having a weekly meal plan, especially during a pandemic, will help significantly reduce your food waste, save money, protect your health, and protect the environment all at the same time. When you create a weekly meal plan, such as the examples provided in Recipe 1, you know exactly how much of each ingredient or food item you need to buy, as well as when you plan to use it. This saves time and money and reduces frustration, as well as making your shopping trips more efficient – all good things (especially during a pandemic when going to a grocery store may increase your anxiety and/or exposure to COVID-19). Creating a weekly meal plan is something that we do in our household.

Another handy tip for expediting your grocery shopping and enhancing your efficiency getting in and out of the store is to have a shopping grid – such as the example in Table 8.5 – that can help you identify where in the store certain items are located. This has helped my family immensely in our grocery shopping. We take our weekly meal plan, look up any necessary recipes, identify the ingredients we will need, write in the amount or number of each item, and then use the grid below to guide us through the store. We make changes to or update the spreadsheet only when items change locations within the store. As an example, Table 8.5 is *our* shopping grid. Items are grouped together by location in the store. This grid works well for us since we frequently rotate many of the same recipes every few weeks and can quickly and easily cross off any items we will not need that week. Creating a table or grid like this for yourselves may help you as well!

Furthermore, by knowing exactly how much of each food item we need, we rarely throw any food away. Lists like this help us to avoid buying foods we will not use or impulse buying foods we do not need or foods that might otherwise end up sitting on a shelf somewhere, eventually spoiling or expiring.

On that note, another major cause of food waste is related to the tiny imprinted "best by" dates found on food packaging. These "best by" dates are a *suggestion* to the consumer, and they refer to the latest date that the product will have the *best* flavor, taste, and texture. "Best by" dates are *not* expiry dates, and depending on what the food is, there may not be much degradation even after the "best by" date.[446,447]

Another date you may frequently see on food packaging is the "sell by" date. The "sell by" date is really aimed at retailers (not the consumer), and it lets them know when the product should be sold or removed from the shelf. However, once again, the "sell by" date does not represent when a food will expire. There are many foods that can be sold after their "sell by" date and still be safely consumed at home. Use your common sense – obviously fresh produce will not last nearly as long as dried beans will after their "sell by" or "best by" dates.[448]

One thing you *can* use these "sell by" and "best by" dates for is storing your foods more methodically. To prevent foods from expiring, spoiling, or losing some of their palatability, you can use the "best by"

Table 8.5 Example grocery shopping grid

Produce	Produce	Refrigerated item	Household items	Pantry staples	Miscellaneous
bananas	mushrooms	tofu	bar soap	chips	corn tortillas
apples	lettuce	grapefruit juice	toothpaste	rice	sandwich bread
tomatoes	bag cabbage	orange juice		dried beans	cereal
avocados	onions	mango juice	**Cleaning items**	pasta	pita chips
basil	garlic	soy yogurt	bleach		
cilantro	potatoes		dish soap		**Snack items**
carrots	limes	almond milk			peanuts
celery	pineapple	unsweet almond	**Baking items**	**Canned/jarred**	hummus
scallions		unsweetened soy milk	whole-wheat flour	artichoke hearts	
peppers	**Shelf-stable**		yeast	tomato paste	**Odds and ends**
iceberg lettuce	cornmeal	**Frozen items**		canned tomatoes	jelly
tomatillos	dried cranberries	frozen berries	**Oil/vinegar**	black beans	peanut butter
fruit	steel-cut oats	frozen corn	balsamic vinegar	apple sauce	honey
mango	lentils	frozen edamame	malt vinegar		
tangerines	split peas	soy cream	olive oil		rolled oatmeal
	white wine				

date to guide you in shelving your foods at home. The way this works is that you store foods in an order that puts the food item closest to expiring in front. This is known as the "FIFO" principle: "First In, First Out." By using the FIFO principle, you will nearly always use the oldest items first and be less likely to throw perfectly good food in the rubbish bin. This will go a long way toward reducing your unintentional food waste.

Another helpful tip is to place perishable food items at eye level in the pantry or refrigerator so that when you open the door they will be the first items you see and the first items you will reach for. Another alternative includes storing foods in ways that will stretch their shelf life and help them last longer. One of these storage methods is freezing certain food items and thawing them only when you are ready to use them. Another long-term storage suggestion is jarring or canning your own foods. For more on how to do that, see Recipe 6.

But we are all human, and despite all of our best efforts, it is still inevitable that some foods will spoil or expire, or their quality will diminish with time and will need to be discarded. When that happens, instead of putting them into the trash or rubbish bin – destined for the landfill, where they will emit methane – put them into a composting bin or waste diversion bin, where they can decompose in a safer and healthier way for the environment (see Recipe 7 for composting).

Finally, to help stop or prevent food waste, you can shop more at farmers markets or at community-supported agriculture (CSA) cooperatives, where the produce may not look as "beautiful" as what you might find in the grocery stores, but it will certainly taste just as good (and probably much better), and it will help reduce the volume of foods that get thrown away or plowed over on farms. You can also encourage your local grocery store to sell "ugly fruits and vegetables" at discounted prices as a way to stem the tide and slow down the global rate of food waste.[449,450] Our local grocery store has a rack in the back of the produce section where they sell "ugly" fruits and vegetables, often at one-third of the price.

These strategies are beneficial to the environment and to the consumer, since "ugly" but unspoiled two-headed potatoes, carrots, or parsnips are still more than safe to eat, and you know what? Once you puree them or chop them up into a stew, a salad, or a juice, you will never remember what they looked like when you bought them.

To sum up Recipe 4, by planning ahead and purchasing only what we will eat, by being more prudent food purchasers and consumers, and by being cognizant of how much food we waste, we really *can* do a lot to protect our health, our wallets, and the planet. And while there will always likely be some food waste, there are more sustainable and environmentally friendly ways we can discard it. For example, to be more environmentally friendly and cost-conscious, we can buy in bulk (Recipe 5), we can freeze, can, or jar excess bulk foods and leftovers for longer-term storage (Recipe 6), and, if there are no other options but to discard foods, we can compost (Recipe 7).

Recipe 5 Buy in Bulk to Reduce Packaging

Walk through any aisle of the grocery store and you will see packaging, packaging, and more packaging. Some of this packaging material is made from paper products, which can often be recycled and refurbished into new paper products. But much of the packaging you see will also contain or be made entirely of plastic.

At least 40 percent of all plastics produced each year (some 161 million tons) goes into packaging materials that are used just one time and then get discarded. Nearly half of all the plastic materials ever produced have been manufactured since 2000.[451] Think about that – nearly half of all plastics that exist today have been made in the last twenty-one years. This is alarming.

And while in the last several years the use of reusable grocery bags has increased, the COVID-19 pandemic has halted this behavior in its tracks. Since the pandemic, many grocery stores have stopped allowing customers to bring in their own reusable bags due to the risk of contamination.[452] This has been a real boon to the plastics industry.

In "normal" times, I would recommend buying foods from "bulk bins" at your local grocery store and using your own personal reusable bags as a way to reduce your use of plastic and single-use packaging. However, amid the COVID-19 pandemic, many grocery stores have significantly reduced or eliminated their use of "bulk bins" as a way to prevent the virus from spreading, which can occur when multiple people handle the same dispensing utensils.**************************

_____

************************** The coronavirus has been shown to remain viable on plastic materials for up to three days.

Given these changes in circumstances during the pandemic, the following are some other ways you can reduce your personal use of single-use packaging:

- Shop at wholesale markets where you can more easily buy in bulk. Wholesale markets tend to sell products in larger sizes (and often use boxes instead of plastic containers for packaging).
  - Along these lines, you can also get groceries delivered to your home from some of these wholesale markets where your personal shopper will deliver a box to your door – filled with your groceries – rather than plastic bags.
- Shop at farmers markets, often located outside in the fresh air, where you can typically still use your own reusable bags.

As briefly mentioned above, wholesale markets will often sell boxes full of fresh fruits and vegetables (as opposed to single units of each). Larger quantities of fresh produce without harmful packaging are more environmentally friendly and often significantly cheaper than purchasing single pieces of fruit or handfuls of vegetables that are then placed into single-use plastic bags.[453]

Farmers markets also tend to sell their products in bulk, and frequently in boxes, paper cartons, or other means that are more environmentally friendly. Often at farmers markets you get fresher and better-tasting produce picked right off the farm – sometimes even that morning – and you reduce the amount of packaging that would have been used by a retail grocer (more on this in Recipe 9).[454]

By reducing the amount of plastic we use only one time and then dispose of, we can stem the flow of plastic materials that end up in the environment, where they pollute waterways, groundwater, and the oceans (see Chapter 4 for a more in-depth discussion). Furthermore, when we buy foods in bulk, we can reduce our use of and exposure to many of the synthetic chemicals – which are not inert and can leach into the foods we eat – that are used in the packaging, storage, and processing of foodstuffs that might be harmful to human health over the long term.[455,456,457]

Outside of "pandemic times" – when we can once again bring our own reusable bags into grocery stores – do consider getting mesh reusable produce bags that are inexpensive and washable and that can be used over and over again, both for bulk purchases and for individual item purchases at a bulk stand. The COVID-19 pandemic has been a particularly challenging time to try to reduce the use of plastic. However, there *are* still ways we can be more mindful about our plastic use.[458]

For example, if you are a single person or have a small family, you can still utilize many of the food shopping tips I have just described – you may just need to be more creative and thoughtful about how you store or preserve your bulk purchases for longer-term use. But there is no need to worry – I give you some tips on how to do just that in Recipe 6.[459]

If you do not live near any wholesale markets, farmers markets, or other bulk-sale markets and your only option is a "typical" grocery store, convenience mart, or bodega, then you may have no other option but to purchase smaller-sized foods (and you cannot beat yourself up over it). If that is the case, you can try to discuss options with the market's manager and inquire as to what their bulk-purchasing powers might be.

Alternatively, there may be other community-supported options available near where you live that you can take part in. For example, you may be able to take part in bulk food delivery or produce box delivery to your home. These types of services frequently offer a variety of box sizes so that you can purchase only what you will eat rather than discard the excess (for more on this, see Recipe 9).†††††††††††††††††††††††††††††††† As mentioned above, if you do buy more than you can eat, there are ways to store, preserve, and use the excess foods that prevent waste. I describe some of these right now in Recipe 6.

Recipe 6 Can, Jar, or Freeze

> Little things make big things happen.
> —John Wooden

I know what you're going to say: "But canning and jarring are things my grandparents did! Why should I do it? I don't even know how to begin!"

Well, let me put you at ease.

---

†††††††††††††††††††††††††††††††† I recommend searching online for "produce delivery box" and seeing what companies are near you. You can frequently obtain healthy, in-season produce from these companies that is cheaper than what you can find at the grocery store. There are several such companies in the United States, Europe, Australia, and elsewhere.

Canning and jarring are methods of preserving foods in airtight containers, such as steel or tin cans or mason jars, to extend their shelf life.[460] A canned product – such as shelf-stable beans – could last as long as five years and still be safe to eat (as long as it is handled and stored properly).[461]

Canning, jarring, and freezing are great ways to preserve foods for a variety of reasons, including:

(1) *They significantly decrease food waste:* In Recipe 4, I discussed how much food we throw away each year on farms, in processing, and in our homes. But what if, instead of discarding the excess foods we purchase, we were to process them and store them for longer-term use? That is what canning and jarring allow us to do.

The process of canning and jarring foods seals in the nutrients contained in those foods and prevents toxic bacteria (in most instances) from harming or degrading the texture, healthfulness, or safety of the food.

Freezing food items can also extend their shelf life and reduce waste. However, for certain types of food products, the shelf life of the frozen item may be quite a bit shorter than the shelf life of the same canned or jarred item. The shorter shelf life is primarily related to the changes that occur in the texture, quality, or scent of that food when it is frozen or when it dries out from "freezer burn."‡‡‡‡‡‡‡‡‡‡‡‡‡‡‡‡‡‡‡‡‡‡‡‡‡‡‡‡‡

As an example, if you purchase an entire flat of strawberries (roughly 20 pounds, or 9 kg) from a farmers market, you know that within two or three days you may be able to eat one-quarter to one-half of the flat and the rest will spoil. So your choice is to buy less of the flat at a more expensive price or to buy the entire flat and store the strawberries in such a way that you can use them at a later time. One option is to freeze some of the strawberries right after you buy them and use them in a smoothie or add them to oatmeal later. Another option is to process some of the strawberries into a jam, which is fairly quick and easy to do – and delicious, I might add – that you can store on the shelf or in the refrigerator in a mason jar for several weeks, months, or even years.

---

‡‡‡‡‡‡‡‡‡‡‡‡‡‡‡‡‡‡‡‡‡‡‡‡‡‡ Worth noting is that freezer burn is less likely and/or is delayed with a frosting freezer that requires occasional defrosting (as opposed to the frost-free freezers that many of us prefer).

By freezing, processing, and/or canning the bulk of the strawberries, you will significantly extend their life, save money, and reduce your food waste. Not only that, but one of the best parts about canning is that you retain the nutrients and the flavors of some of the best-tasting, seasonal, and most nutritious foods you can buy, and you get to enjoy their flavors and healthy benefits into the future, at no extra cost to you.[462] This goes for many types of fruits and vegetables and is a great way to enjoy the "fruits of your labor" for longer.

(2) *They buffer you against food shortages:* In addition to extending the life of the foods you buy, canning, jarring, and freezing can give you a bit of a buffer against food shortages, such as those we experienced (in some places) during the COVID-19 pandemic.

Just imagine opening up your pantry doors and seeing shelves filled with jarred or canned foods that you can rely upon rather than worrying about food shortages at your local grocery store. Similarly, when your freezer is filled with a variety of frozen fruits, vegetables, and prepared foods (that you rotate in and out on a regular basis), you will not have to worry as much about shortages of frozen items at the store; you will have your own cache.§§§§§§§§§§§§§§§§§§§§§§§§§§§§§§§

For individuals or families who may be food insecure or who live paycheck to paycheck, buying in bulk – which is usually cheaper – and finding ways to store those foods long term can go a very long way toward improving food security.

(3) *They increase your access to fruits, vegetables, and other nutritious foods when they might otherwise not be available (or if they are out of season):* In some locations, such as at higher latitudes, it can be more difficult to find or afford fresh fruits and vegetables during certain times or seasons of the year. If you want to get the benefits of produce all year long, then you may need to be able to process and preserve these items for year-round consumption.

While the last several decades have seen the mass globalization of foods – meaning foods can travel halfway around the world on planes, ships, or other vehicles – this doesn't mean that the foods we purchase off-season will taste as good, nor will they retain all of their nutrients. Moreover, in our efforts to be more environmentally

§§§§§§§§§§§§§§§§§§§§§§§§§§§§§§§ Even months after the pandemic began, freezer shelves and canned foods are still at times sparse at my local grocery store.

friendly and significantly cut our "carbon foodprint" – or the miles our food has traveled to get to our plates – we can store and enjoy foods all year long that are locally grown and seasonal using these food storage methods.

As discussed at length in Recipe 1, plant-based foods, including those that are frozen, canned, or jarred, benefit health significantly, and being able to have them on a daily basis, even off-season, is extremely important for your overall health and well-being.

Just like finding certain types of foods all year can be difficult in cold climates, the COVID-19 pandemic affected food supply chains, making it more difficult and more expensive to obtain certain types of foods. Therefore, having a cache of easily accessible, ready-to-go frozen, canned, or jarred food may be just what the doctor – or dietitian – ordered.

These reasons among many are why canning and jarring and other methods of food preservation are so valuable (and also why they have been in use for so long).

(4) *Once you have made the initial investment, canning and jarring is a relatively inexpensive way to store and extend the life of foods:* While there is an initial investment for canning and jarring supplies – such as mason jars and lids – often these are relatively inexpensive when bought in bulk. Additionally, once you have bought the jars, they are reusable indefinitely (well, until they break). Lids and screw tops do need to be replaced each time, but they are relatively inexpensive, especially considering how much can be saved on food!

Freezing foods, meanwhile, requires simply having a working freezer and some good-quality airtight containers that are either freezer-safe or made out of a durable, reusable plastic.*************************** If you happen to have good-quality glass products that are stackable and can withstand being frozen (and thawed), these can be an even better choice, as you can reheat food items in them without the fear of having plastic chemicals leach into your food.

Therefore, in an effort to be as healthy, cost-effective, sustainable, environmentally friendly, and beneficial to your overall well-being as possible, I highly recommend using long-established preservation

*************************** I only make this note on using plastic as I am fairly certain that many people reading this book have sturdy Tupperware or a similar product in their pantry, and this is a perfect use for it – better than tossing it in the trash.

methods, such as canning, jarring, and freezing, as ways to extend the shelf life of the foods you buy (and/or grow).††††††††††††††††††††††††††††† Give it a go – it's a lot of fun!

However, I will mention that these methods are not 100 percent reliable at maintaining the quality, taste, or safety of foods – nothing is. Even with the best of intentions and the best methods, discards will still happen. Therefore, the next best action we can take is to compost our food discards instead of throwing them into the rubbish pile.

Recipe 7 Compost

The more I have come to understand how composting benefits the environment, the more excited I am to be able to address it here in this recipe.

As a primer, composting is a natural process involving the "rotting" or decomposition of organic matter.‡‡‡‡‡‡‡‡‡‡‡‡‡‡‡‡‡‡‡‡‡‡‡‡‡‡ Crop residues, animal waste, food discards, and other fibrous materials including paper are all candidates for the composting process. The decomposition of these materials occurs using microorganisms under controlled conditions in the presence of oxygen.[463]

When we use composted materials and mix them back into soil in farming applications, a wide range of nutrients get recycled. Furthermore, instead of contributing to the waste heap, composted materials produce a stable, rich, and highly productive organic fertilizer. By reusing these nutrients, we can significantly improve "soil, plant, water, human, and planetary health" using what are known as regenerative agricultural practices.[367],§§§§§§§§§§§§§§§§§§§§§§§§§§§§§§

In addition to acting as an organic fertilizer, the use of compost also improves a soil's resilience to stressors such as drought, disease, and toxicity by:[367]

---

††††††††††††††††††††††††††††† For a complete guide to home canning, see https://nchfp.uga.edu/publications/publications_usda.html.
‡‡‡‡‡‡‡‡‡‡‡‡‡‡‡‡‡‡‡‡‡‡‡‡‡‡ The term "organic" in this context refers to carbon-based, natural compounds such as plant materials, manures, compost, wood shavings, and food discards – not the methods by which foods are grown.
§§§§§§§§§§§§§§§§§§§§§§§§§§§§§ Regenerative agricultural practices enrich soil, improve water uptake and storage capacity, increase biological diversity, and enhance ecosystem services. Regenerative agriculture also captures carbon that would otherwise be emitted into the atmosphere from landfill processes and keeps it inside the soil, where it can be used to feed the next generation of crops and even reverse climate change and lower the amount of carbon dioxide in the atmosphere.

- *Increasing the ability of the soil to absorb more water:* Compost contains many microorganisms and organic matter, which hold on to carbon as well as water. Organic matter holds ten times its weight in water; therefore, for every 1-percent increase in soil organic matter, an acre of land will absorb and store an additional 25,000 gallons (94,600 liters) of water.[464] That capability is becoming increasingly important as more land suffers from desiccation and desertification under intense agricultural practices (and drought conditions).

- *Increasing the soil's resistance to disease:* There are a number of organisms found in soil that have the potential to be plant pathogens. These include fungi, bacteria, viruses, nematodes, and protozoans.[465] Fungi cause the majority of plant diseases found in agricultural fields. On conventional farms, fungi are dealt with (read: killed) through the heavy use of fungicides. Insecticides, bactericides, and other chemicals are also frequently applied on farms to kill other types of pathogenic organisms.

   Compost, on the other hand, contains a range of carefully managed (read: beneficial) microorganisms, which thrive on oxygen and organic substrates that, when added to soil, crowd out and compete with the harmful or pathogenic bacteria, viruses, or fungi that might cause disease or harm to crops.[465] This means that compost can improve the health of soil and manage microorganisms all on its own without the need for added chemical killers.

- *Decreasing the risk of soil contaminants and toxicity:* Finally, contaminants and pollution can come from a range of natural or human-made sources and can build up and exceed the maximum permissible levels allowed in natural soil environments.[466] If soil becomes contaminated with potentially toxic elements, including arsenic, cadmium, chromium, copper, lead, and others, this can have very serious health effects on the animals and humans that consume foods grown in that soil.***************************

   Recent studies demonstrate that there are many organic materials in compost that have the ability to bind to,

*************************** A nationwide soil quality survey in China conducted by China's Ministry of Environmental Production found that 16 percent of all soil samples contained levels of contaminants that exceeded recommended standards. Moreover, there are certain villages with higher-than-average and higher-than-expected levels of cancer that have been associated with soil pollution in China (Palansooriya et al., 2020).

break down, or restrain a number of soil contaminants or toxins, and that this may potentially render them effectively harmless.[466],†††††††††††††††††††††††††††††††††††

Another benefit of compost is that it increases the uptake of healthy nutrients – such as vitamins, minerals, and other bioactive compounds – into the crops themselves. This is an important recycling of nutrients that occurs due to microbial activities within the compost itself.

Taken together, the use of compost significantly moderates environmental threats, frequently produces higher yields (at a much lower cost), improves the nutrient profile of the crops themselves, and decreases – if not eliminates – the need for inorganic fertilizers, pesticides, herbicides, and other chemical inputs.[463]

There Are Even More Benefits to Composting!

In addition to all the benefits just described, there are several other great reasons to compost. Compost is a gift that just keeps on giving, and one of the most important benefits of compost is that the act of composting itself breaks down organic "waste" materials into beneficial end products, including:

- *Carbon dioxide instead of methane:* While carbon dioxide does not sound like a positive, let me explain. Since composting is an aerobic (oxygen-requiring) process, carbon dioxide is produced as the end product rather than methane (which would be the end product if the waste were instead diverted to landfill). As discussed earlier in this book, methane is far worse for the atmosphere than carbon dioxide. Furthermore, much of the carbon dioxide that *is* produced during the composting process gets stored within the compost itself rather than being released into the atmosphere. This makes compost a net carbon-negative product that actually pulls carbon down and out of the atmosphere.[467]
- *Nitrogen-rich humus and moisture:* The actions of the microorganisms within compost, along with the generation of water, heat, and ammonia (an end product), produce a rich, dark-colored, and stable

††††††††††††††††††††††††††††††††††† This is primarily accomplished through a reduction in leaching of the contaminant or a reduction in phytoavailability (plant uptake) of the contaminant.

organic humus, which acts as a nitrogen-rich fertilizer when applied to or mixed into soil.[463] Adding humus into soil can save farmers significant amounts of money by reducing their need for inorganic fertilizer and water (remember, every 1-percent increase in organic carbon in the soil increases the soil's water storage capacity by 25,000 gallons per acre).[463] This is huge!

- *Fertilizer and micronutrients:* Each year, the amount of synthetic fertilizer used on conventional farms increases. Since the 1950s, the use of fertilizers around the world has increased by fifteen to twenty times, and it will continue to do so in order to keep up with the global demand for food.[367] Yet compost can significantly reduce or even eliminate the need for synthetic fertilizer by naturally increasing nitrogen levels in the soil. Richer soils can grow more crops, and having more crops can keep soil from washing out or running into streams and waterways. This is in contrast to what happens with very dry, tilled, and denuded soils.[367]

  In addition to higher levels of nitrogen, compost also recycles vitamins and minerals, as well as potassium and phosphorus, among other important organic compounds, giving crops higher levels of healthy nutrients.[468,469,‡‡‡‡‡‡‡‡‡‡‡‡‡‡‡‡‡‡‡‡‡‡‡‡‡‡‡‡‡‡‡‡] This means that, calorie for calorie, foods grown with compost in the soil contain significantly more nutrients than foods grown "conventionally" with synthetic nitrogen fertilizer.[467]

On a large scale, compost can save billions of dollars every year on farming inputs and water, produce more nutritious foods at a lower cost, enhance health and nutritional well-being, and protect the environment all at the same time. If more industrial farms, community plots, individuals (on their lawns), and cities (in parks) were to use and advocate for the use of composted materials, hundreds of millions of tons of solid organic waste and food waste could be diverted away from landfills every year.[367,470] The following is one example of how this could happen:

> In 2005, the then-Mayor of San Francisco, Gavin Newsom (now the current Governor of California), introduced a program called "Zero Waste." The program works by

‡‡‡‡‡‡‡‡‡‡‡‡‡‡‡‡‡‡‡‡‡‡‡‡‡‡‡‡ Some inorganic fertilizers are derived from petroleum by-products.

incentivizing consumers' use of the compost and recycling waste streams and disincentivizing the use of the landfill waste stream.

Each household or apartment building receives three containers from the city – one for each of the three waste streams. When households fill their compost or recycling bins, they pay nothing; but when they fill their landfill bins, they pay a lot to have it collected and dealt with.

This program is quite effective, and over 650 tons of compostable materials are collected each day in San Francisco.[367] Once these materials are turned into compost, the compost is sold to farms, orchards, or vineyards near San Francisco, making this an excellent business plan.

This truly is a case of "one man's trash is another man's treasure."

Since so much food waste occurs in our homes (as described in Recipe 4), sending it to be composted rather than to landfill can save nearly 1.3 billion tons (1.3 gigatons) of methane gas emissions every year.[471],§§§§§§§§§§§§§§§§§§§§§§§§§§§§§§§ This represents a 30-percent reduction in emissions normally produced from this sector, equivalent to every person reducing their personal emissions by 0.2–0.5 tons every year.[472]

This one action has a farther-reaching impact on reducing greenhouse gas emissions than does changing all of the lights in your home from incandescent light bulbs to light-emitting diodes (LEDs).*************************** It both reduces emissions and turns soil into a carbon sink: this is a win–win in so many ways.[466,467]

Compost also reduces the risk that food-borne pathogens will be released into the environment or pollute downstream watersheds – which can occur more readily in a landfill.[473,474] Pathogens and microbial organisms that may be present in food or organic waste die from

---

§§§§§§§§§§§§§§§§§§§§§§§§§§§§§§ Remember, 40 percent of all foods in the United States are discarded, and as much as one-third of all foods grown throughout the world end up in landfill.

*************************** If you change all of the light bulbs in your home from incandescent light bulbs to LEDs, you would save less than 0.1 tons of carbon per person per year – much less than switching to a plant-based diet (1 ton of carbon per person per year) or composting all of your food discards (0.2–0.5 tons per person per year).

the heat that is generated during the composting process.[473] This is yet another way to keep the food supply safe.

What's more, the use of compost can help farmers lower their costs and improve the nutrient profiles of the foods they grow. This can potentially benefit you – the consumer – and translate into lower prices *and* healthier foods.[475,476]

On that note, if you have your own personal garden – in your yard, in a raised bed, or even a miniature one on your balcony (as we do) or other small outdoor space – you can take these same composting principles and apply them yourself. Instead of throwing your food waste into the rubbish or trash bin, compost them on your own and liberally apply the humus end product onto your own garden. This will save you water and fertilizer and help the environment simultaneously.[477]

If you do not have your own garden, you *can* still compost, and you *can* still contribute to this recipe by using your municipal compost bins, which fortunately seem to be increasing in number all the time. And if that still is not an option – based on where you live – you can ask your local grocer whether or not they compost and see if you can contribute to their heap.

If none of these options exist where you live, however, there may still be other potential opportunities to engage in composting, including with community gardens and CSA cooperatives, both of which may accept your food waste or compost scraps.[478] For more on community gardens, see Recipe 8.

Recipe 8 How to: Home and Community Gardens

For those of you who are lucky enough to have your own backyard garden, a raised bed or planter, or a small plot of land, having the ability to grow some of your own herbs, fruits, and vegetables is fun, healthful, and a truly sustainable endeavor. As an added benefit, the foods you grow usually taste better (and have more flavor) than what you can buy in the store.

Even in a relatively small area, roughly 4 feet × 10 feet (1.2 m × 3.0 m), you can grow quite a variety of produce, such as herbs, lettuce, brussels sprouts, carrots, celery, kale, peas, corn, and other vegetables and fruits (especially melons). You can mix and match to create your own favorite garden depending on what you like to eat, and you can even include some heirloom varietals,††††††††††††††††††††††††††††††† such as tomatoes, spinach, certain lettuces, lima beans, heirloom squash and melons, carrots, and peppers, to name a few.

If you are like my family and I, however, and you live in an apartment or condominium building that does not have a communal plot of land and all you have is a small balcony, you can *still* have a small garden within reach to grow some items that can be used in a homemade soup, salad, or main dish. On our small balcony, which is only 12 feet (3.6 m) long, we grow two different types of tomato plants, basil, carrots, spinach, and other fresh herbs.

From our balcony's bounty we have made fresh pesto (cheese-free) and tomato sauce and have contributed greens and herbs to our salads. It is always nice to add a final touch to a meal with fresh herbs right from your own garden – a truly local garden.

Neighbors of ours have grown cucumbers, peppers, and other herbs on their balconies and have shared their bounty with us, as we have with them. If you grow too much of one thing, you can swap it for something else with your neighbors and start your own small community garden.

We have also grown dwarf orange and lime trees on our balcony, planted in whiskey barrels, which were needed to achieve an appropriate depth for the roots of the trees. While these trees did not produce much fruit (they were only a few years old), they were still fun to grow and provided a touch of nature to our home. When the trees grew too big for our balcony, we moved them to our local community garden.

If you do not have access to a backyard, a planter, or balcony, however, do consider looking into a community or shared garden near your home, where you can share a plot, typically a raised bed, with some friends or others in your community to grow some of your own produce. We are very fortunate to have joined our local community garden, situated at the University of California, Los Angeles (UCLA), where we work. Community gardens are also a fun way to meet people and contribute to society.

---

†††††††††††††††††††††††††††††††††† "Heirloom" denotes a traditional variety of plant not associated with large-scale commercial agriculture. The following website provides a decent number of suggested heirloom varietals: https://gardenerspath.com/plants/vegetables/heirloom-vegetables-garden.

At the UCLA community garden, we share a 4 foot × 10 foot (1.2 m × 3.0 m) plot with another UCLA family who we did not previously know. We have been amazed by the amount of produce we have grown in this small plot and how much of it we have used to create new recipes. Figure 8.10 shows our community garden plot right after we harvested a fair amount of the radishes and arugula.

In our community garden planter, we have grown carrots, radishes, arugula, tomatoes, pumpkins, watermelon, corn, green beans, kale, squash, beets, and several other lettuce greens, along with a handful of other fruit and vegetable varieties. In fact, we grew so much arugula – which grows like a weed (I am absolutely serious about that) – that we could not eat it fast enough! So, of course, rather than discard what we could not eat, we gave some to our plot partners from the community garden and some to the campus food closet – for those in need – and we threw the stem trimmings back into the garden so that they could degrade into compost to feed the next generation of seedlings.

Growing some of your own foods gives you the opportunity to *engage* with your food systems more intimately and gives you a better understanding and appreciation of what goes into producing the foods you eat. Too many of us are so far removed from the food systems that feed us – except when we go to the grocery store and look for the most beautiful specimens – that we have forgotten what it takes to feed a community, let alone a country, or the entire world.

Growing your own foods also gives you a first taste of the freshest, most flavorful "farm-to-table" food because there is no trans-portation time.[479] Sometimes we have simply skipped the table! I cannot tell you how many times we have picked fresh arugula from our garden and thrown it into a quick salad on the spot (we rinsed it first, of course). There is nothing quite like it.

In addition to a better understanding of food-growing systems, growing some of our own food has given our young son a hands-on opportunity to learn about biodynamic soils,‡‡‡‡‡‡‡‡‡‡‡‡‡‡‡‡‡‡‡‡‡‡‡‡‡‡‡‡‡‡‡ composting, worms, water, and the importance of ladybugs (ladybirds

---

‡‡‡‡‡‡‡‡‡‡‡‡‡‡‡‡‡‡‡‡‡‡‡‡‡‡‡‡ "Biodynamic soils" refers to how soil properties change with time, external changes, water, temperature, etc. (http://soilquality.org/basics/inherent_dynamic .html).

Figure 8.10 Our plot at the University of California, Los Angeles community garden

in the UK; a natural "pesticide" in the fight against pests such as aphids).

There is evidence that engaging with a community garden and growing some of your own food encourages you to eat a more healthful and sustainable diet than you might otherwise be able to afford and on a more consistent basis.[480] These benefits seem to be related to improving the accessibility, affordability, and availability of these foods within the community, and this is especially the case for lower-income communities.[481,482]

More than 80 percent of individuals in the United States live in cities. More than 75 percent of Europeans live in cities, and across the globe, more than half of all people live in urban environments where relating to the land and growing your own food is difficult, if not impossible.

High-rise buildings and apartment buildings further exacerbate this "distance from the land," as they rarely provide shared pieces of land that can be used by the community to grow foods. And while there are some planned communities that may set aside a communal space for this purpose, they are still lamentably rare overall.[483]

Having access to this wonderful community garden, nearly in our own backyard, was *our* introduction into urban and peri-urban agriculture. A similar garden can be your entrée, too. More and more cities are offering their citizens the opportunity to grow some of their own foods through classes and workshops at cooperatives and community gardens.[484],§§§§§§§§§§§§§§§§§§§§§§§§§§§§§§§§§

Community gardens also provide a model for sustainable urban living by contributing to and facilitating local engagement with food systems, by increasing access to locally grown foods, and by bridging the gap from farm to table.[485]

Community gardens (as well as home gardens) also help the environment and our own health in several ways:

- First, community gardens increase our access to fresh fruits and vegetables, which we all need for optimal health.
- Second, community gardens reduce the amount of agricultural resources needed to feed a community because of the generous use of compost and close monitoring of the garden by its community members.
- Third, as was the case during the COVID-19 pandemic, community gardens may eliminate or reduce the need for supply chains that may

§§§§§§§§§§§§§§§§§§§§§§§§§§§§§§§§§ You can find your own local community garden at www.communitygarden.org.

become hampered by outside forces and therefore may not function as they normally do.[486]

- Fourth, community gardens reduce food-related transport emissions by nearly eliminating the distance food must travel from the farm to the home.

The benefits of community gardens may be especially important in developing countries, where community gardens significantly shorten the supply chain and improve food security by giving rural communities a way to cope with drought, erratic rainfall, and variable food prices.[487]

Furthermore, active engagement with a community garden supports a plant-based diet by improving access to and the affordability of fruits and vegetables – this is often related to the discount or "payment" given in the form of sweat equity.[488] Yet another benefit of working in a community garden is its positive effect on mental health by providing opportunities to commune with nature and by increasing our sense of self-efficacy as we grow, produce, harvest, and consume our own fruits and vegetables.[489]

Due to the space limitations of a community or home garden, however, you may not be able to grow the full complement of fresh produce that you might find in a large grocery store. But remember, some of the produce in your grocery store may have traveled thousands of miles to get there, may be out of season, and may not have nearly as much flavor as what you can grow in your own garden seasonally at different times of the year. Moreover, your own garden will be more naturally resilient to pests as a result of having a more varied array of produce than what is grown in a typical monoculture farm (which consequently often requires large amounts of pesticide).[490]

For those of you who do not have access to your own personal garden or a community garden, keep your eye out for farmers markets and produce stands that are popping up more and more around cities and towns (even during the COVID-19 pandemic). These are also great ways to bring farm-fresh produce closer to home, as are CSA cooperatives (see Recipe 9).[491]

If you are looking to start your own backyard or community garden, Table 8.6 highlights some potential crops to grow. (Please note that these specific crops may vary by the time of year and also by where you live. There may also be some seasonal overlap. I recommend researching what crops grow well in your area.)

Table 8.6 Potential crops to grow in a home or small community garden

Summer	Fall	Winter	Spring
Cucumbers	Pumpkin	Beets	Melons
Lettuce	Winter squash	Potatoes	Cucumbers
Arugula	Onions	Carrots	Beets
Peas	Parsnip	Turnips	Carrots
Spinach	Carrots	Garlic	Tomatoes
Zucchini	Leek	Celery	Beans
Corn	Artichoke	Onions/shallots	Broccoli
Eggplant	Parsley	Asparagus	Potatoes/sweet potatoes
Peppers	Parsnips	Peas	Raspberries
Herbs (e.g. basil)	Chives	Broad beans	Onions

Some of the ideas in this table are derived from the US Department of Agriculture Community Garden Guide[492] and some of them are derived from my own experiences.

Table 8.7 Potential crops to grow on a balcony

Summer	Fall	Winter	Spring
Cucumbers	Mint	Garlic	Tomatoes
Peppers	Cilantro	Beets	Cucumbers
Tomatoes	Basil	Onions/shallots	Raspberries
Summery squash	Radishes	Peas	Basil
Celery	Parsnips	Carrots	Cilantro
Lettuce greens	Chives	Herbs	Zucchini

Some of the ideas in this table are derived from the US Department of Agriculture Community Garden Guide[492] and some of them are derived from my own experiences.

And if you only have a balcony available to grow a few plants and herbs, Table 8.7 provides some recommendations for what you can grow.

And, of course, if you do not have a yard, a balcony, or any other place where you can grow your own fruits, vegetables, or herbs, you can always look to farmers markets or CSA (see Recipe 9)!

Recipe 9 When You Cannot Grow It Yourself, Shop at Local Farmers Markets and Use Community-Supported Agriculture!

One of the nice things about farmers markets is that they give you the chance to meet some farmers who are local to your area and who grow some of the foods that you may eat. Farmers markets also give you an opportunity to ask farmers about some of their farming techniques, the variety of crops they grow, and how far away they are located. You may even be surprised by what you learn, and hopefully in a good way!

All over the world, there are farmers markets and produce markets selling products that are locally grown or produced.* Where I live in Los Angeles, California, there are several weekly farmers markets spread out throughout the city. Many of these markets were even allowed to stay open during the COVID-19 pandemic.

In my experience, the farmers and other sellers who participate in these markets really enjoy talking about their products and the efforts they put into producing them. Many of the farms represented in these markets are certified organic, and some may even use regenerative farming techniques. Those that are not yet certified organic have often adopted many organic practices and are on their way toward certification.[493]

In one study conducted in the United States looking at 1,016 farms, 45 percent used only conventional practices, 19 percent used a mix of conventional and organic practices, and 36 percent used only organic practices. Yet, of the 36 percent that used only organic practices, 71 percent chose not to certify as organic, often due to the cost of certification or confusion over the process.[494]

In Europe, organic farming has grown by 70 percent in the past 10 years. As of 2017, 14 percent of Spain's, 13 percent of Italy's, 12 percent of France's, 9 percent of Germany's, and up to 28 percent of other countries' agricultural land is organically farmed, increasing both the prevalence and availability of organically grown foods.[495] Some of these foods are sold in local farmers markets, where they were picked that morning at their freshest and ripest, giving the consumer the most nutrients and flavor possible.[479]

---

* I highly recommend using Google or another application to find local farmers markets.

Too often, produce sold at grocery stores is picked days if not weeks in advance of when it is actually sold at market. This often is because the produce is grown in other parts of the world, shipped to a warehouse, and then finally distributed weeks later. By the time you finally take the produce home, it has lost much of its flavor.[496] So although there may be more variety of produce in grocery stores, it may have less flavor and vibrancy and fewer nutrients.

This leads me to another major benefit of farmers markets: less packaging. I can think of many examples where I have seen prepackaged or prewrapped produce in the grocery store. I've even seen bunches of bananas sold in plastic bags or shrink-wrapped on a Styrofoam tray. I've seen bell peppers wrapped in plastic or prepackaged in plastic bags. I've seen bunches of grapes sold in plastic bags and handfuls of oranges sealed inside plastic containers. There is just so much plastic everywhere! And most of it gets used once and then discarded in the trash (see Chapter 4 for more discussion on plastic pollution).

In contrast, at most farmers markets I have been to around the world, from North America, to Europe, to Australia, packaging – if any is used at all – is reserved for keeping the produce safe during travel, and it is used, reused, and then reused again by the farmers.

So why not do the environment a favor – in many ways – by stopping and shopping at *your* local farmers market when you next need fruits and vegetables. You will be:

(1) Helping your local farm economy and local farmers;
(2) Reducing the distance your produce traveled to get to your plate;
(3) Reducing the amount of packaging used, especially single-use packaging;
(4) Encouraging your local farm to use organic farming methods (and perhaps regenerative farming methods as well).

If, however, there are no farmers markets within a short walk, bicycle ride, or drive of you, consider participating in CSA. CSA is a "growing social movement that endeavors to make direct connections between the producers of food and those who consume it."[497]

The CSA concept originated in Europe and Japan. In the United States, the first CSA farm began in 1985 in Massachusetts, and within fourteen years, there were an estimated 1,000 of them.[497]

In its simplest form, CSA is a contract between a farm and a group of consumers (shareholders, members, or subscribers) giving the farmers an idea of how much to produce. CSA also provides farmers with up-front capital that they use to purchase necessary farming inputs.[497]

While the majority of items produced by a CSA farm are organic fruits and vegetables, other products that may also be distributed include meat, poultry, dairy products, cider, honey, flowers, and even maple syrup. The members or shareholders of a specific CSA farm will typically receive products on a weekly, biweekly, or monthly basis depending on their chosen subscription.[498]

There are many benefits of participating in CSA, including:

(1) Knowing you are contributing to and helping the local farm economy and local farmers.
(2) Feeling more connected to farmers and a community, as many CSA farms engage in social events and work details for members to help at certain times of the year.[499]
(3) Reducing single-use plastic packaging as most CSA products are delivered in recyclable cardboard boxes or wooden crates.[500]
(4) Cost savings on often organic produce.
(5) An increased sense of well-being and self-efficacy. This is usually borne out of consuming healthy, recently picked produce and addressing the broader goals of ecological sustainability and health.[499,501]
(6) Your food travels fewer miles from farm to plate and has a lower environmental footprint.
(7) Farms involved with CSA typically practice more sustainable, ecological, or regenerative farming methods.[502]
(8) The foods received from a CSA farm taste better than store-bought foods.

Finally, "CSAs are seen as entities [which] empower people by providing an arena in which they can act on their principles and against the values of the global food system."[499]

Another important aspect of CSA is that it has the ability to improve and increase access to healthy foods for lower-income households who might not otherwise be able to afford or consume some of these healthier food products.[503] Also worth noting is that, during the COVID-19 pandemic, CSA membership increased as a result of people preferring delivered food options.[504]

The reasons to participate in or join a CSA farm are many and varied, as are the types of products that are available to you, which often shift with the seasons.

My recommendation, as a practicing dietitian and as someone who cares deeply about both your health and the environment, is this: If you are unable to grow your own fresh produce or visit a farmers market, please consider finding a CSA farm near you that fits your budget and offers produce that you will enjoy. You may find yourself a little bit healthier and possibly even a little bit happier. For many, food is an emotional and cultural experience, and good food grown well can brighten your day – go on, give it a try!

Another great benefit of shopping at farmers markets and CSA farms is that most of the food products come without plastic wrapping, which is great when we are all trying to reduce, reuse, and refuse plastic.

Recipe 10   Reduce, Reuse, and Refuse the Plastic: There Are Better Alternatives!

In Chapter 4, I addressed several of the detrimental effects plastics can have on health and the environment. Here, I give suggestions for what we – as individuals and consumers – can do about it.

As mentioned in Chapter 4, each of us is responsible for using a huge amount of single-use plastic products every year, with only a fraction getting recycled.[165] Each year, every one of us uses enough products wrapped or encased in plastic – made from petroleum – to drive a car several tens of miles; remember, just twelve single-use plastic bags are made from enough oil to drive a standard car one mile.[505]

Many days of the week, many of us stop at coffee shops for a single-use, wax-lined cup that we cover with a single-use plastic lid and perhaps drink with a single-use plastic straw. Some of us may even take one of those little single-use plastic plungers designed to prevent coffee from spilling out of the small hole at the top of the lid. As you can see, doing that day in and day out uses a *lot* of plastic.

Many of us are also familiar with taking the plastic bag that is offered to us at the grocery store, convenience store, mall, or restaurant, whether we need it or not, even if we are just walking 120 feet to our car.

Some regions of the world are better at curbing their plastic use than others. For example, the populations of Denmark and Finland have reduced plastic bag use to just four per person per year (these are pre-COVID-19 numbers), and the people of Ireland have reduced their use of single-use plastic bags from 328 per person per year in 2002 to just 18 per person per year in 2017.[506] This amount of progress is simply amazing.

Some European countries have banned all plastic bags except for those that are biodegradable and compostable. Other European countries have levied a fee for plastic bags, which has cut their use in half.[506,507] Some US states, including California, Connecticut, Delaware, Hawaii, New York, Oregon, and Vermont, have banned single-use plastic bags at large retail stores or have begun to levy fees; California, my home state, was the first to do so in 2014.[508] There are also some US cities – including Boston, Chicago, Seattle, Boulder, New York, and Washington, DC – that have passed plastic bag bans, even though their states have not yet passed such bans.[508] On a similar note, Vermont, Maine, and New Jersey have banned polystyrene (Styrofoam) use, and Maryland, Colorado, and Oregon have similar bans in the works.[509]

As mentioned earlier, much of this progress has been halted or reversed by the COVID-19 pandemic, and we again are seeing an increase in the use of single-use plastics. One reason for this increase is the need for personal protective equipment (PPE) such as masks, gloves, and other medical equipment *made from* plastics. Another reason is that when many restaurants were forced to close to in-person dining and to switch exclusively to takeout or delivery, the use of takeout containers – often made from plastic or polystyrene – sky-rocketed. Yet another reason for this increase is that grocery stores stopped allowing people to bring in their own reusable shopping bags as the virus can live on various materials (for varying lengths of time).[510] These changes related to the COVID-19 pandemic poten-tially increase the risk that plastic bag bans could be reversed long term or perhaps even permanently – to the detriment of the environment.

Compounding matters, many recycling and municipal waste services became more limited during the pandemic. This made it

Figure 8.11 Plastic rubbish on a beach in Southern California

increasingly difficult to process the exponential increase in PPE and other plastics, and allowed even more of it to end up in landfill and in our oceans.[510]†

It is estimated that between 5 and 12 million metric tonnes (5.5–13.2 million tons) of plastics enter the oceans every year. Plastics of all sizes and shapes make up roughly 75 percent of all marine litter (see Figure 8.11). Plastics have many negative health effects on oceanic wildlife (and the seafood many of us eat), as previously addressed in Chapter 4.[511,512]

By 2050, plastic waste in the oceans may weigh more than all of the fish remaining in the oceans.[513] The models that predict this are based on the length of time it takes for plastic to degrade in the oceans, the exponential increase in plastic use over the last fifty years, the fact that we are adding more plastic to oceans every day, and the fact that we do not reuse or recycle nearly as much plastic as we could or should.[514] We are converting this planet into one gigantic garbage

† www.weforum.org/agenda/2020/08/disposable-masks-plastic-pollution-coronavirus-covid-19

disposal – it sometimes feels like we are living in the opening scenes of the Disney film *WALL-E*.

All of this plastic in the oceans not only looks bad; it also causes harm to our health, to the health of our children, and to the health of the environment and many ocean-dwelling species (many of which we depend on) in all sorts of ways (see Table 4.1). If we carry on as we are, the amount of plastic that is entering the oceans and the environment is only set to increase, creating more pollution and wreaking more havoc, until we stop or drastically cut back on our use of plastic. At the same time, we need to increase our reuse and recycling of those plastics that already exist.

Plastic has become a mainstay in our lives, and there will likely always be a use for it for some products – PPE, face shields, medical equipment, or certain packaging – until a more environmentally friendly equivalent can be substituted. However, there is also a need to find a more efficient solution for reusing and recycling, as it will likely be difficult (if not impossible) to entirely remove plastics from our lives. Yet there are areas in our lives where safer alternatives and more sustainable options may be used to stem the flow of plastic, even during a global pandemic.

Before I provide examples of some potential substitutions, please take a few minutes to consider how often and how much plastic you use in your own lives. I urge you to run through the checklist in Table 8.8 to help put you in a frame of mind to consider some alternatives.

When you pause and review some of your own plastic use (since this list is not exhaustive, you may have thought of additional areas), you may find that you use far more plastic than you actually need to. Wherever you are on the continuum of plastic use, there are things we all can do.

There are many companies creating a number of great alternatives to single-use plastic products that are better for the environment and that can be used over and over and over again. By supporting them, you can support environmentally friendly businesses, choose ethical consumerism, and lead by example.

To make an even bigger and more lasting impact, it is important to remember also *to refuse* the plastic, even when it is handed to you. It's easy – just say, "No, thank you."

Table 8.8 Plastic use questionnaire

Question	✓
Do you take a plastic bag from a convenience store or restaurant just because it was handed to you with the item you purchased when you did not really need it?	
Do you use single-use plastic or wax-lined cups or coffee lids when you could have gone without or could have brought your own?	
Think about your last grocery shop: Did you leave the store with a cart filled with ten, fifteen, or even twenty single-use plastic bags?	
Do you buy individual pieces or small amounts of produce that are prewrapped in plastic packaging?	
When you pick produce at the market, even a single item, do you use separate plastic bags?	
If you buy a single-use plastic bottle of water, does the cashier place it in a plastic bag so that it is easier to carry even though you are about to open it and drink from it?	
When you take home leftovers or takeout from a restaurant, do you take it home in a Styrofoam or plastic container placed inside of a plastic bag?	
Do you recycle your plastic bags?	
Do you reuse your plastic bags or repurpose them into trash- or rubbish-can liners?	
Do you use small bottles of shampoo, conditioner, or soap when substitute bar forms are available?	
Do you buy individual snack items that are prepackaged? (For example, some companies prepackage peanuts, trail mix, or apple slices into single-serve portions in small plastic containers inside of a larger plastic bag.)	
Do you buy bottled water when you could be drinking tap water or be refilling a reusable bottle from drinking water taps?	

The following are some examples of ways to reduce your use of single-use plastic:

Reusable bags: Instead of taking single-use plastic bags from the store, bring your own reusable totes, canvas bags, nylon bags, or mesh bags. Even if you are not permitted to bring your own bags into the store, you can still purchase your items, place them back into the cart – unbagged – and then bag them yourself once you have gotten to your car.

We have been doing just that since the start of the COVID-19 pandemic, since stores will not bag our items for us in our own reusable bags. A positive outcome of this is that bagging your own groceries might even help you burn off a few of those "COVID-19 pounds" you may have gained during the pandemic.

There are many reusable bags available today that fold up nice and small so that you can easily take them with you and always have them at the ready. I frequently put a collapsible, reusable bag in my coat pocket so that if I come upon a farmers market while out for a walk or decide to pick up a few items at the store I do not need to take a single-use plastic bag from the store. Similarly, when you drive to the store, keep your reusable bags in the car so that you always have them ready.

I have received many compliments on my bags, even from complete strangers, as many of them have pictures of my son or my dog on them, and some of my bags even have pictures of both. I have been asked numerous times where I have my bags made and am always happy to pass on the information. For just a few dollars you can create your own reusable bag, and you too can look chic while caring for the environment.[‡]

Coffee/tea cups: Another area where we can use less plastic is our coffee and tea habits. There are many of us who go to coffee or tea shops and spend significant amounts of money on coffee or tea served in wax-lined cups that come with a plastic lid, neither of which are usually recyclable.

More than 16 billion disposable coffee cups and lids are used every year.[515] In order to make all of these cups, 6.5 million trees are cut down, 5 billion gallons of water are used, and huge amounts of energy are expended – enough to power 54,000 homes in developed countries.[516]

Instead of using all these resources, save money – as much as $1,000 per year – and save paper and plastic by brewing your own coffee or tea at home. With a small amount of effort, you can make yourself a better drink than your high-end coffee shop.[517,§,**]

[‡] I happen to use Shutterfly to make my reusable bags, but you can also use photo labs at pharmacies or other online photo sites that have this offering.
[§] I strongly recommend seeking out organic and fair trade coffee and tea companies.
[**] Keurig cups, coffee pods, or Nespresso capsules are not environmentally friendly – they are individually wrapped (often in plastic) coffee products that make only one cup at a time. You would be better off purchasing whole beans, ground beans, or granules.

However, if you do want go to a coffee shop and/or support a local business, bring your own mug or sealable carafe. Many coffee shops offer discounts to customers who bring their own, so you can still get your favorite coffee at your neighborhood coffee shop and save some money in a way that is better for the environment![††]

Water bottles: More than 480 billion single-use plastic bottles were sold worldwide in 2016. This is up from the roughly 300 billion bottles sold per year only one decade prior.[515] That is a lot of bottles – over 60 per person per year! Instead of buying single-use plastic water bottles, look into getting sturdy, reusable water bottles made from reinforced glass or metal. There are many companies now that produce and sell beautiful and reusable water bottles. Some of these bottles will keep your beverages at a lovely temperature (read: they keep cold drinks cold and hot drinks hot).

Another incentive for using reusable water bottles is that bottled water sold in single-use plastic is very expensive. Liter for liter, or gallon for gallon, bottled water costs thousands of times more than tap water.[‡‡] Moreover, water from a single-use plastic bottle (or even a multiuse plastic bottle) can expose you to plastic chemicals and/or contaminants that leach out of the bottle and into the water.[518,519,520]

The answer is clear: Invest in a stainless steel or reinforced glass bottle. This small investment can last many years and save you a lot of money.[§§]

Many cities around the world have water-refilling fountains where you can safely and cleanly refill your reusable water bottle free of charge. Paris now even has refilling stations for sparking water!

I have had the same stainless steel water bottle for over ten years (and it shows), and I take it with me whenever and wherever I travel. It has saved me a great deal on water costs, keeps me safe from plastic chemicals, and is easily accessible. It has been a great investment – especially for my health, as drinking enough clean water is vital for optimal health.

[††] Policies may have changed as a result of the COVID-19 pandemic.
[‡‡] One bottle of water could cost as much as $3–5 (or more!) for a liter or two (a quart or two), whereas the same amount of water from the tap will cost only a fraction of a penny (<0.04 cents).
[§§] Instead of spending $450 (this would be using store-brand water bottles) or as much as $1,100 (this would be more like using expensive Fiji water) per year (or more) on bottled water, you could spend $20 on a reusable bottle and drink as much water as you want!

Straws: Around the world, more than half a billion plastic straws are used each year.[515] The vast majority of people do not need to drink their beverage with a straw (there are, of course, some medical conditions that require this, but these are few and far between), and plastic straws are one of the more insidious plastic products as they cannot be recycled, they are small, and they often end up in waterways, on beaches, and in the ocean.[521],***

Today, there are many alternatives to plastic straws, including paper straws, hay straws, wheat stem straws, grass straws, agave straws, metal straws, and silicone straws, and the best of all is the no-straw straw, where you pick up the glass and drink directly from it. So next time you order a beverage, consider bringing your own reusable straw, asking if the seller has a nonplastic alternative straw, or just drink straight from the glass.

Packaging: Over half of the plastic that was thrown out in 2015 – more than 141 million metric tonnes (155 million tons) – came from packaging. Packaging is one of the major contributors to plastic pollution.[515]

While it may be difficult to influence the packaging practices of large companies that ship products to each other (although employees and certainly CEOs at those companies can advocate for change), we as individuals all have a choice in what types of products we buy, the types of packaging they come in, and how we spend our money.

As mentioned in Recipe 5, when you buy in bulk, you reduce the amount of packaging you take home and have to do something with.

When you get takeout or take home foods from a restaurant, inquire as to whether you may bring your own container to reduce the amount of packaging the restaurant uses: they may very well appreciate the cost savings at their end.

When making online purchases, ask for the fewest number of shipments possible to limit your packaging waste and, where possible, request plastic-free packaging.

Contact your local stores or markets and request that they use plastic-free packaging or compostable products. The more people who ask, the more companies will feel compelled to provide these.

*** I have seen far too many images of birds, whales, or turtles with straws stuck in their noses, stomachs, or other orifices.

There are, of course, facets of life where plastic is inevitable and unavoidable. But if we all pay attention to our own use and to how foods and other items are packaged or served, we can make a big difference in the amount of plastic that gets discarded, especially when large numbers of individuals get involved, and especially at a local level, where we may have more opportunity to act as a positive influence.

If nothing else, ask your coffee shop to switch over to more sustainable options. There are now companies that produce compostable and biodegradable coffee lids that will fit standard coffee cups – and prices are coming down all the time as demand grows. Be the agent of change. Reach out to corporate companies and ask them about their sustainability practices.

Before you vote, look into a politician's voting record and see whether or not they support and vote for regulations and laws that require companies to invest in environmentally sustainable practices and products.

More and more, city by city, state by state, country by country, and region by region, changes are happening, and we can all do our part to support those changes. By assessing our own use of plastic and how pervasive it has become in our own lives, we can make a difference, and we can conscientiously make changes, for when all of our changes are added together, they make a real difference.

We can also participate in beach or river cleanup events or even create our own. We have more power than we think we do, and leading by example is important.

So here are some suggestions for what *you* can do right now:

(1) Try to stay away from plastic. Don't use it, don't buy it, don't cook with it, and don't drink out of it. But, if you must, you can ...

(2) Reuse it, repurpose it, recycle it, and keep it from getting into the oceans.

(3) Purchase glass or stainless steel storage containers instead of plastic ones.

(4) Purchase reusable mesh (or cotton) produce bags so that at the grocery store or farmers market you can put produce into these bags as opposed to single-use plastic bags (that rip half of the time anyway).

(5) Skip the plastic lid on a single-use coffee cup, request a compostable lid, or, better yet, bring your own reusable mug or carafe – you may even get a discount.

(6) Or save yourself some money and make your own coffee at home with real coffee, grounds, or instant granules.[†††]

(7) Jar your own vegetables (see Recipe 6) or buy jarred vegetables instead of canned to avoid BPA- or bisphenol S (BPS)-lined cans. This is particularly important with acidic foods (e.g. tomatoes), as acidic foods increase the amount of BPA/BPS that can leach into the food from the lining of the can.

(8) Take your own reusable bags with you whenever you are out – you never know when you will make a purchase. Or skip the bag entirely if you can carry or wheel your items to your car or if you only have a short distance to walk.

(9) Try out bar shampoos, bar conditioners, bar soaps, or other cosmetic products that are not in plastic containers.[‡‡‡]

(10) Eat less seafood or avoid it altogether, as one of the larger sources of plastic in the oceans comes from lost or discarded fishing gear, which is made from plastics. By avoiding seafood, fewer ships will be fishing the oceans and leaving plastic rubbish behind.

(11) Overall, be mindful of all the plastic that is in your life and do what you can – starting today – to cut back on or cut out your use of plastic. Sometimes all we need is one small change to really get the ball rolling.

We inherited a cleaner Earth from our parents and we want to leave it better for our children, not worse. We need to keep the land and the oceans as clean as possible, not have them be filled with plastic and harmful chemicals. We can choose more eco-friendly products and we can be more informed about these issues. That's where change begins – with knowledge. For more on compostable and eco-friendly alternatives, see Recipe 11.

Recipe 11 Seek Out Compostable and Eco-Friendly Products

Unless someone like you cares a whole awful lot, nothing is going to get better. It's not.
—Dr. Seuss's *The Lorax*

[†††] There are some really great organic, fair trade instant coffee granules available on the market now – my favorite one is Mount Hagen organic and fair trade instant coffee.
[‡‡‡] Simply search for these types of products online to determine where you can find/purchase them.

Now that I have discussed the many detrimental effects of plastic on our health and the environment, I want to present some ideas for seeking out compostable and eco-friendly replacements.

Nowadays, there are many companies that produce biodegradable products, which include biodegradable plastics and bio-based plastics that are made from synthetic or natural polymers. The type of polymer produced depends on the process or function that the end product will be used for.[522]

Biodegradable plastics were introduced as materials designed to contribute to environmental conservation by completely breaking down into carbon dioxide and water – end products that are much less detrimental to the environment than traditional plastics that persist within the environment.

Biodegradable plastics break down through the actions of naturally occurring microorganisms – such as bacteria, fungi, and algae – on the materials themselves (much like with composting).[523] The time that is needed to decompose biodegradable plastics depends on the materials' makeup, the environmental conditions – such as temperature and moisture – and the products' location (e.g. on land, in water, or in an industrial facility).[524]

One type of biodegradable plastic – compostable plastic – rapidly turns into humus, the end product of composting that forms the basis of soil in a natural environment (for more on composting and humus, see Recipe 7).

Not all biodegradable plastics are compostable, however.[525] Only products that meet the American Society for Testing and Materials (ASTM) specifications D6400 and D8686 or the European Committee for Standardization (CEN) specification EN 13432 are both biodegradable and compostable on land. The ASTM is an international standards organization that develops and publishes consensus technical standards for a wide range of materials, products, systems, and services.[526],§§§

Biologically based plastics, also known as bioplastics, are produced from natural, renewable plant sources (rather than from

---

§§§ Like the ASTM, there are also the CEN, the International Organization for Standardization (ISO), and the German Institute for Standardization (DIN). Each of these organizations specify the biodegradation, disintegration, and eco-toxicity criteria that must be met for a plastic to be called compostable.

petroleum by-products). The most common bioplastics are made from starch, cellulose, hemicellulose, lignin, or plant oils, which are extracted from plant and wood biomass.[523],**** Although many bio-based plastics are incinerated after use, some can be recycled. Bio-based plastics are considered environmentally friendly, even when burned, as the carbon dioxide generated can be used to grow more biomass through photosynthesis.[527]

The use of biodegradable and bio-based plastics has grown rapidly over recent years as replacements for wood, metal, glass, and petroleum-based plastics due to their advantages over these materials – including being tougher, lighter weight, corrosion resistant, and their ease of processing into various colors, shapes, and designs.[528]

While there are differences between biodegradable plastics (those that break down in the environment) and bio-based plastics (which are made from plant biomass, but also break down in the environment), both are welcome substitutes in a world where plastics have become so ubiquitous.

The following are some examples of bio-based plastics:

Polyhydroxyalkanoate (PHA) is a bio-based plastic produced naturally by bacteria and genetically modified plants.[523] There are some companies that are also trying to make PHA from food waste.[529] Polyhydroxybutyrate (PHB) is a type of PHA that may also be used; however, PHB is very expensive, as only small quantities are currently produced by bacteria.

PHA can be used as food wraps, water bottles, cups, plates, and coatings for paper and cardboard, and it also has many medical uses, including in sutures, gauze, and coatings for medicines. PHA can also be used in place of many oil-based plastics currently in use, such as PE, PS, PVC, and PET (refer back to Table 4.1 for more details on these plastics).[530]

PHA can be blended with starch or cellulose to make a less expensive product. PHA on its own may be five times more expensive than its oil-based counterparts. However, as technologies improve, the cost of producing PHA will go down.

One benefit to a PHA blend is that it decomposes more quickly than straight PHA. In fact, a PHA blend will completely decompose

---

**** Biomass is a combination of the words "biology" (life) and "mass" (quantity), and it represents organic resources such as wood or plant materials.

within two months in a soil-rich environment.[531] Like any plastic, however, PHA decomposes significantly slower in a marine environment than it does on land. But, given enough time, unlike its oil-based counterparts, PHA *will decompose entirely* in a marine environment.[532]

Polylactic acid (PLA) is a plastic substitute made from bacteria-fermented plant starch – usually corn – but it can also be made from sugarcane, sweet potatoes, and other plants that produce glucose.[528] PLA is a long chain of lactic acid molecules, and it is relatively inexpensive to produce as compared to PHA. However, PLA can be brittle, and its use may be more restricted than it is for PHA.

PLA can be used as an alternative to oil-based plastic in products such as compostable packaging, including clamshell carryout food containers. PLA can also be used in biodegradable flatware and in some medical products, including sutures, which will dissolve in the body within ninety days.[532] PLA can also be used in 3D printing and is completely carbon neutral.[533]

Although PLA does biodegrade, it does so more slowly than PHA (taking anywhere from six to twelve months) and does not compost easily in soil. Instead, PLA can be diverted to a controlled composting environment in an industrial composting facility where it is heated to $140°F$ ($60°C$) and fed a steady amount of digestive microbes.[527]

PLA decomposes into carbon dioxide and water (as opposed to methane or petroleum by-products). Unfortunately, though, like any rubbish, if PLA products are not properly diverted to an industrial composting facility and are instead sent to landfill, they will take far longer to biodegrade, and they may degrade into methane gas, which is significantly more detrimental to the environment than carbon dioxide.[534,535,††††]

PLA does not decompose well in marine environments and may be a physical hazard (or danger) to marine animals who accidentally consume it. However, PLA is still a better alternative to oil-based plastics (which also are physical hazards to marine animals) for several reasons:

(1) PLA is nontoxic and has a better material safety profile.
(2) It comes from renewable resources.

†††† The problem with landfills is that there is too little light, not enough heat, and too little oxygen to appropriately decompose PLA.

(3) It is carbon neutral.

(4) It uses 65 percent less energy to produce and form into products than conventional plastics.

(5) It allows us to leave more fossil fuels in the ground.[536]

Cellulose acetate (CA) is another bio-based plastic. CA is a product derived from the cellulose chains of plants, primarily cotton, wood, and crop waste. CA can be molded into solid plastics, coatings, and photographic films and filters, and it can be used to make cellophane, a biodegradable cellulose-based film.[537]

Evidence suggests that CA will degrade by 70 percent in nature over an eighteen-month period. It can, however, be industrially composted or recycled much more quickly.[538]

Starches are yet another biomass that can be made into plastic materials when they are treated with heat, water, and plasticizers to produce thermoplastics.[‡‡‡‡] Major sources of starch include maize (corn), wheat, potato, and cassava. Starch-based plastics can be used to make packaging, bags, agricultural mulch films, tableware, and flowerpots, and they can also be molded into food packaging and consumer goods.[539] Starch-based plastics are often used as eco-friendly alternatives to polystyrene (PS; Styrofoam), and starch can even be added to bio-based and conventional plastics to make them more biodegradable.

While concerns have been voiced that transitioning food crops and/or cropland toward bio-based plastics may have negative impacts on global food security, similar transitions have already taken place to produce biofuels. The most common crops that are used to make biofuels include soybean, corn, canola, and other food-based oils. These are the same crops that are used to make bio-based plastics. Therefore, the case has been made that using crops or land to produce bio-based plastics will not represent a significant increase over those currently used to produce biofuels.[527,540]

Furthermore, if regenerative farming techniques are used to grow these crops (as described in other recipes in this chapter), they could in fact represent an enormous environmental boon and be used to

---

‡‡‡‡ Plasticizers are substances – typically solvents – added to a resin to produce or promote plasticity and flexibility and to reduce brittleness. Some examples include phthalates and phthalate derivatives.

Table 8.9 Bio-based plastic types, uses, and biodegradability

Bioplastic	What it is used for	Biodegradability/compostability
PHA	Food wraps, cups, plates, paper coatings, sutures, gauze, medicinal coatings	2 months in soil
PLA	Grocery bags, bottles, cups, food packaging, plates, medical sutures, 3D printing	1–6 months in commercial facilities 6–12 months in soil Dissolves in 90 days in acidic environments
CA	Solid plastics, coatings, photographic films/filters, cellophane	Up to 70 percent over 18 months, up to 60 percent in marine environments
Starches	Food packaging, bags, tableware, consumer goods, agricultural mulch	If "compostable," requires 90 days in industrial facilities If "biodegradable," requires 100 days for 46 percent to degrade and 2 years to fully degrade

sequester carbon.[367] If done successfully, these crops, among others, could allow bio-based fuels and plastics to be net carbon negative.

Table 8.9 gives a quick summary and overview of the bio-based plastics just described.

In addition to bio-based plastics, I will now discuss some examples of biodegradable plastics that are on the market.

Examples of Biodegradable Plastics

Polybutylene succinate (PBS) is a material that can be produced either from fossil fuels or from biologically based materials, including corn, cassava, and sugarcane.[541] PBS can also be combined with other bio-based polymers or fibers to improve its quality.[532] PBS is most frequently used to make food packaging, tableware, agricultural mulch sheets, plant pots, and hygiene products such as diapers. It can also be used to make fishing nets.[532] PBS decomposes into the end products water and carbon dioxide when under the right conditions and with the right microorganisms in the soil.

Polycaprolactone (PCL) is a synthetic polyester that can be used to make compostable bags as well as in medical applications to make sutures and fibers, surface coatings, adhesives for shoes and leather, and as stiffeners for shoes and orthopedic splints. PCL will decompose in the presence of certain yeasts, and more than 90 percent of films and 40 percent of foams made from this material can degrade within 15 days.[542] PCL is easily blended with starches and lignans and has rubber-like characteristics, making it a durable plastic material. PCL will degrade under a variety of conditions, including in soil, in water, in sewage, and in compost.[543]

Polybutyrate adipate terephthalate (PBAT) is a polymer produced from fossil fuel derivatives and is sometimes combined with starch to increase its biodegradability. There are companies that are trying to produce this polymer from renewable or bio-based sources. PBAT is used as a substitute for low-density polyethylene (LDPE) and high-density polyethylene (HDPE) and is used to make garbage bags, wrapping films, disposable packaging, and tableware such as cups and dishes. It is not only biodegradable, but it is also compostable.[544] PBAT can biodegrade through the enzymatic action of microorganisms such as bacteria, fungi, and algae present in the natural environment. PBAT can also biodegrade nonenzymatically in temperature-controlled industrial composting conditions.[545]

Polyvinyl alcohol (PVOH/PVA) is a resin that can be used in packaging films to replace LDPE and HDPE. It can also be used as a coating or additive for paper and board products. PVOH can be used in food packaging as it does not produce any toxic or harmful by-products when incinerated or digested. It has also been used in detergents, pharmaceuticals, agricultural chemicals, and as pods for dishwashing and laundry soaps.[546]

Table 8.10 provides a quick summary and overview of the biodegradable plastics just described.

Every one of these biodegradable plastics degrades within three months in an industrial composting facility or within one to two years in soil, landfill, or marine environments.[532,§§§§]

[§§§§] An important thing to remember with biodegradable plastics and bioplastics is that in order for them to be significantly better for the environment than standard plastic materials, they must be properly disposed of in recycling containers, through industrial compost diversion, or be biodegraded appropriately. If they are not, they will pollute the environment and may cause physical harm to animals that happen upon them.

Table 8.10 Biodegradable plastic types, uses, and biodegradability

Bioplastic	What it is used for	Biodegradability/compostability
PBS	Food packaging, service-ware, agricultural mulch sheets, plant pots, hygiene products	3 months in an industrial composting facility or 1–2 years in soil or landfills
PCL	Compostable bags, sutures, surface coatings, shoe adhesives	Gets decomposed by yeast (up to 90 percent in 15 days)
PBAT	Garbage bags, wrapping films, disposable packaging, cups and dishes	3 months in an industrial composting facility or 1–2 years with enzymatic bacteria, fungi, and algae in the natural environment
PVOH/ PVA	Packaging films, paper and board coatings, may be used in detergents or pharmaceuticals	Is often incinerated or enzymatically digested

As of 2015, bio-based and biodegradable plastics accounted for only 1 percent of the world's total plastics production, but this may grow significantly in the coming years.[547] Bio-based plastics such as bio-PE and bio-PET are identical to their petroleum-based counterparts and can be used in the exact same applications, but without the same toxic chemical properties and with less harmful consequences to the marine environment and animals.[547]

Although bio-based and biodegradable plastics are not perfect or risk-free alternatives, they are still better and safer options for our health and the environment compared to the plastics made from petroleum.

In addition to being better for the environment (when properly used and disposed of), there is evidence to suggest that when individuals use environmentally friendly products and make eco-friendly purchases, they feel a greater level of social responsibility, self-efficacy, and empowerment. Moreover, when individuals engage in ethical consumerism – the intentional purchasing of products and services considered to be made ethically, with minimal harm to or exploitation of humans, animals, or the natural environment – they are also more likely to make environmentally friendly and sustainable purchases in the future. This, of course, brings me neatly to the next recipe.[548]

Recipe 12 Engage in Sustainable or Ecological Tourism

> We can have environmental justice and positive economic
> development that generates jobs for communities, but at the same time
> fosters a cleaner environment. They are not mutually exclusive.
> —Damu Smith

Ecotourism is defined as "responsible travel to natural areas that conserves the environment, sustains the well-being of the local people, and involves interpretation and education. Ecotourism unites conservation, local communities, and sustainable travel by:

- Minimizing physical, social, behavioral, and psychological impacts on the environment;
- Building environmental and cultural awareness and respect;
- Providing direct financial benefits for conservation;
- Designing, constructing, and operating low-impact facilities; and
- Recognizing the rights and spiritual beliefs of indigenous people of the community and working with them to produce empowerment."[549],*****

In contrast, consumption-based tourism tends to stress local lands, increase pollution, exacerbate habitat loss, deplete local natural resources, and increase pressure on endangered species because of human overconsumption, and frequently this happens where resources are already scarce.[550]

It is worth noting that *all types* of tourism produce more than 5 percent of total global greenhouse gas emissions, and the transportation that is associated with tourism accounts for 90 percent of this.[550]

Roughly 7 percent of all tourism today is ecotourism, but this is growing. Ecotourism represents a set of morals – or a philosophy – for how tourism can minimize negative effects and maximize positive conservation outcomes for local people, animals, their habitat, and the planet.[551]

Ecotourism tends to contrast starkly from conventional, consumption-based tourism in its use of resources, and it is a kinder and more environmentally mindful way to enjoy yourself and your surroundings.[551]

***** TIES – The International Ecotourism Society (https://ecotourism.org/what-is-ecotourism).

As I addressed throughout Chapter 6, there are too many "tourist attractions" around the world that involve the maltreatment and abuse of animals, many of which were stolen from the wild and who have been put through harsh techniques to tame them of their wildness and "train" them to do tricks for the "amusement" of humans.[†††††]

With ecotourism, however, instead of taking from nature or taking an animal from the wild and making it fit a particular image, you instead observe and understand nature and animals in their natural habitats for what they already are and what they inherently offer to this Earth (not to humans). Ecotourism works to maintain and perhaps even improve the environment through conservation, care, and respect of nature and its inherent beauty and intricacy.

Studies suggest that ecotourism and spending time in nature improves feelings of psychological well-being and satisfaction (much more so than consumption-based tourism).[552] Furthermore, there are longer-term benefits associated with engaging in ecotourism, including a demonstrable increase in environmentally responsible behaviors even after going home.[553]

Moreover, spending time in nature in the presence of wildlife has been shown to remove the socially constructed confines of time and space, allowing participants to become completely absorbed in the experience and to be entirely in the moment. This has been shown to awaken a deep sense of well-being as a result of wonderment and awe that spans beyond the encounter and leads to a myriad of psychological health benefits and a recognition that conserving habitats and wildlife is necessary to the future well-being of the planet and, frankly, ourselves.[554] This is in stark contrast (again) to consumption-based tourism or tourism based on human-made attractions, which do not elicit nearly as positive a response.[555]

Given the emotional, physical, and social benefits of engaging with nature and ecotourism for individuals, communities, habitats, and the environment, it seems that this type of tourism ought to be the way of the future. Let us strive to do better for our own emotional and psychological well-being and for the well-being of the planet and its species.

[†††††] This can typically be seen in elephant-show or elephant-ride tourism and some zoos or animal amusement parks.

On a similar note, much research has been done that finds similar positive benefits from buying fewer things, having less "stuff," and engaging more in experiences.

Recipe 13 Buy Fewer Things and Less "Stuff"

How much do we really need to make us happy? How many things? How much stuff?

My guess is – for the majority of us in the developed world – a lot less than we may think.

In the last thirty years, a number of studies have been conducted assessing levels of subjective well-being. Subjective well-being is a measure of how positively or negatively a person experiences their own life – essentially, how happy they are.

Many studies of subjective well-being note that higher levels of consumption are, at best, only very weakly related to higher levels of happiness.[556] This means that once basic needs are met – such as having enough food, water, shelter, clothing, healthcare, safety and security – all of the other "stuff" we add to our lives does not increase our happiness. While some of these items may increase short-term pleasure, the consensus is that they do not have any lasting impact on overall well-being or happiness.[556,557]

In fact, studies of happiness (in developed countries) report that only 2–5 percent of an individual's overall "happiness" can be attributed to their income or their consumer behaviors.[‡‡‡‡‡] This means that the other 95–98 percent of a person's perceived happiness is related to other things.

There is strong evidence suggesting that helping others, doing things for others, and spending time with others *are* associated with increased well-being and happiness. Moreover, these studies show that experiences and a person's memory of experiences increase their level of subjective well-being far more than the consumption of material goods.[558]

Another interesting finding is that the longer it has been since an experience, the more that we remember the joy or happiness we felt

---

‡‡‡‡‡ There is one exception to this: For the extremely poor – individuals living at or below the poverty line – incomes, and especially increases in income, are associated with increased levels of happiness.

from the experience (rather than any of the negative feelings of the experience). We tend to remember the good, not the bad. This type of selective memory reinforces the happiness we derive from experiences, irrespective of when it is being measured.[559,560]

It is suggested that the positive feelings we get from an experience last significantly longer than the fleeting positive feelings we get from spending money on new gadgets, toys, cars, or clothing. One reason that has been identified to explain this phenomenon is that experiences are more closely associated with identity, connection, and prosocial behaviors.[560]

Moreover, many "things" we buy have a limited life span and end up in the rubbish pile within a few months to a few years, leaving little to no lasting impression on us other than perhaps the psychological distress of their price tag or the fact that we are throwing them away.[561,562] In fact, studies show that individuals who buy fewer things but engage in more experiences may be happier than individuals who engage in environmentally friendly consumer behaviors such as buying ecofriendly products.[563,§§§§§] This may be explained by the fact that even environmentally friendly products often eventually end up in the rubbish heap.

Therefore, I have to ask: Why *do* so many of us buy more things or more stuff? What type of happiness are we chasing?

While this is anecdotal, I must say that some of the happiest, most joyful, most kindhearted, and most warm people I have ever had the privilege to spend time with were those who have the fewest things. My time in Ethiopia reminds me of this. The individuals I met literally had nothing but the few clothes they wore on their bodies. But despite all of their hardships and despite their extreme poverty, they were extremely generous with their time, their feelings, and their hearts.

What these individuals wanted most in life was not more "things." What they wanted most in life was education, either for themselves, for their brothers or sisters, or for their sons or daughters. What's perhaps even more shocking is that what little they had they offered to share with me, someone who has far more than them. It was

---

§§§§§ This definitely does not mean that we shouldn't buy environmentally friendly items, it just means that when we buy fewer things in general, we tend to be happier, and it *also* is environmentally friendly to buy less stuff.

one of the most touching experiences I have ever had, and we can all learn so much from this.

Rather than chase the newest style or the newest electronics model, save your money for experiences, for ecotourism, for the betterment of your health, and for the environment.

Every time you buy something new, remember that it requires multiple inputs and resources to manufacture and get to market. Instead of always purchasing new, consider buying items that may have been gently used or repurposed, and instead of tossing out your old items, consider donating them to those in need.

By conserving just a bit more we can reduce the number of inputs and resources that are needed, whether they be natural fibers, the water that goes into growing them, or the petroleum by-products that get transformed into plastics that cause environmental damage.

By buying fewer things and less stuff, by being more mindful about what we do with our old and used items, and by turning our attention more toward experiences, we can help the environment, save a bit of money, and perhaps be just a little bit – or maybe even a lot – happier.

Recipe 14 Engage with Your State and Local Officials

No matter where you live, whether in the United States, Europe, Australia, Asia, or Africa, you likely have local government officials and leaders who are responsible for local matters such as public health, parks and playing fields, educational services, local environmental issues, and waste disposal, among other community services. Very often, the individuals who serve these communities, regions, or states come from these communities and are elected by popular vote, and as such are accountable to the people who elected them: their constituents.

Therefore, if you want your voice to be heard on issues related to the environment and sustainability, it is imperative that you reach out to your local officials.

There are cities around the world at the forefront of progressive environmental policies, and often city officials are more effective than state or federal governments at implementing and enforcing some of these policies.

Public participation – often with local grassroots organizations – also helps shape and carry out many sustainability measures.

Compliance with environmental programs and their goals is often better when there is engagement within a community and where there are intersections among public health, public safety, and the environment.[564]

Air pollution and particulates are associated with asthma, chronic obstructive lung disease, lung cancer, congestive heart failure, and cardiovascular diseases. Improving air quality – often done at the city, county, or state level – has a myriad of benefits for health, including lowering direct medical costs, decreasing or avoiding pain and suffering, and increasing the productivity of the workforce due to them experiencing fewer sick days.[565] Therefore, it may be imperative to work with your local leaders if you want to improve your air quality.

Another area where local-level engagement is important – if not crucial – is in the use of plastic bags. There are many municipalities around the world where fees are levied for the continued use of plastic bags. This has led to significant declines in their overall use. Municipal-level fees and enforcement of these fees seem to have done considerably more to change consumer behavior than straight bans alone.[566]

While there are many decisions made solely at the legislative (or government) level, there are also times when issues – especially those related to the environment and sustainability – end up on the ballot for the will of the people to decide. Either way, whether you vote on a specific initiative or for a specific person, taking an interest in and engaging with your local and state officials on these issues will increase their salience.

Depending on where you live, many local officials now have "contact us" web pages, email addresses, phone numbers, message boards, and social media accounts. Additionally, elected officials have offices with staff who are required to listen to you or read your letters.

If you still feel your concerns are not being heard or addressed, advocacy groups represent yet another way to raise awareness of the issues you care about. Advocacy groups typically have regional leaders. Engage with them and other organizations that are actively working on the issues you are concerned about. And if there are no organizations already in place, consider starting your own or engaging with friends or colleagues to increase the prominence of these issues.

As just one example, I work at an academic medical center on a university campus that has a robust sustainability presence and has made many efforts to encourage people to recycle, compost, and discard their waste appropriately – especially in the cafeterias. Yet I noticed that many medical personnel buy their lunch in the cafeteria and take it up to the floors of the hospital to eat in the break rooms with their colleagues, where there were no composting containers and few if any recycling containers (as there were in the cafeteria itself). Moreover, since health-care workers typically only have thirty minutes to one hour to eat their lunch, they do not have the time (or the motivation) to bring their waste back down to the cafeteria for proper disposal.

While personally I *did* carry my waste back down to the cafe-teria every day on my way out, I realized that not everyone was going to be as conscientious! Sometimes you have to meet people where they are, compromise, change what is available, change the defaults, and move the needle through simple but meaningful shifts.[567] So this is what I did.

I took my concerns to the medical center's sustainability man-ager and discussed what I saw as a major opportunity to improve waste management on the floors of the hospital. I worked directly with "deci-sion-makers" on changing the status quo.

After several logistical discussions with infection control, envir-onmental services, nursing directors, and upper management, I am pleased to report that there currently are composting and recycling containers in all break rooms in the hospital and medical clinics on the campus, and now over 38,000 employees can more easily access appropriate waste streams.

While this is not a state-wide, county-wide, or even city-wide change, it *is* a change nonetheless, and it is a change that affects the behaviors of 38,000 individuals, making it easier for them to contribute to environmental sustainability. Moreover, those 38,000 individuals may adopt some of these good practices at home when they interact with their own partners, kids, friends, family, etc. And so one change may carry on up to a bigger scale.

If everyone who reads this book attempted a similar challenge within their own companies or communities, no matter how big or small, you could impact many people in many places.

Very often we hear about the importance of voting and how voting makes all the difference. And while this is a true statement (see Recipe 15), sometimes it is also necessary to tackle a project at the grassroots level, where it is smaller and more manageable, so that we can see that it is possible to effect immediate action and behavior change.

This is actually one of the reasons why I wrote this book – because the ideas in this book are things we all can do, either on our own or in collaboration with others. When we observe and achieve success, it motivates us to work toward achieving bigger goals and making bigger changes on a larger scale – but you have to start somewhere.

> One of the things I learned when I was negotiating was that until I changed myself, I could not change others. —Nelson Mandela

I quote this statement here because I believe it contains an important lesson. It shows that if we want others to do better and to be better, then we must be willing to do these things ourselves first and lead by example. I encourage you to find an area of sustainability that you care about and work toward making a change. It could be as simple as picking up ten pieces of rubbish every time you leave your house (and recycling them appropriately!) or talking to your local government officer about your concerns. It is also important to vote, of course. But even small actions can make a difference. They add up.

Recipe 15 Vote for and Elect Public Officials Who Believe in and Support Climate Change and Eco-Friendly Policies

> Vote as if your life depends upon it, because it does.
> —Justin Dart

There are many individuals around the world, including politicians, who are "antienvironmental," who do not believe in science, who do not believe in climate change, and who are actively rolling back sensible environmental policies and regulations.

When it comes to these beliefs, I have often wondered about those in political power: Do they *really not* believe in climate change? Or is it that they care more about their (and/or their friends') financial gains and political favors in the here and now, such that they are willing to sacrifice the environment for their own children and grandchildren?

I am not to certain we can ever truly know the real answer to this question.

Therefore, it is absolutely imperative that, whenever your city, your county, your state, or your country holds an election, you use your power – your obligation – to vote. Voting is a proxy for your voice, and with your vote, you can vote for individuals who *do* believe in climate change, sustainable policies, and environmental and species protection.

In order to know who to vote for, look at an individual's track record. What are they saying about the environment? About climate change? How long have they been saying it? Perhaps even more importantly, what have they done or are they doing about the environment? If they have a political or legislative background, what bills or laws have they introduced, sponsored, or voted for? What protections have they advocated for?

Climate change and its dangers, including the threat of extinctions around the world, are not to be taken lightly. Climate change affects everyone, no matter where you live. No one is immune, regardless of how much money you have (although perhaps having more money may shield you from the worst effects of climate change for a longer period of time). Everyone will be forced to live with higher temperatures, changes to the land, loss of soil, or loss of farmland; nature is bigger than any one person, city, or country.

Large population centers in parts of China, the United States, Japan, the Netherlands, Thailand, Indonesia, Italy, India, Brazil, France, Bangladesh, and certain islands throughout the world are less than 5 meters (16 feet) above sea level.[568] Rising sea levels as a result of climate change will displace millions if not billions of people in these low-lying areas. Climate migrants – individuals who are forced to leave their homes to seek refuge in other places or countries – may also increase the risk of conflict over limited space and resources.[569] We are already seeing this.

Those who are most at risk from the negative effects of climate change are those who have the fewest resources, the lowest incomes, the fewest safety nets, and the least effective governance.[570] Low-income populations are disproportionately overburdened by hazardous waste sites, dirty air, and contaminated drinking water, with few resources to mitigate or prevent the negative consequences of these threats.[571]

In many places around the world, in the last few years we have witnessed the deregulation of environmental protections, an increase in deforestation, increased threats to endangered species, the dismantling of environmental regulatory agencies, an increased push for dirty energy sources such as coal and oil, and decreased funding for renewable energy sources to such a degree that one could say that the environmental movement has been pushed back several decades.[572] Hopefully this changes soon – for the better.

There are many corporations looking to profit from environmental deregulation, whether by exploring the Arctic for oil reserves, by fishing, mining, and drilling in marine sanctuaries, or by poaching and deforesting terrestrial refuges. What's more, in many places politicians are not only permitting these things to happen, but are even openly encouraging them through executive actions or legislative votes.[573,574]

The only way to change the discourse around these issues – to protect the environment – is to vote for leaders at all levels of government who will defend the environment, who understand the ramifications of pillaging the Earth, and who recognize that sustainable living and climate change mitigation and adaptation are absolutely required. While we will not all vote for the Green Party, in a choice between two candidates, one will always be the "better" candidate for the environment.

Many climate models demonstrate the potential for a "hothouse Earth," where global temperatures could rise by as much as 9°F (5°C), sea levels could rise by 20–30 feet (6–9 meters), where there could be a complete loss of the world's coral reefs and the Amazon rain forest, and where large swaths of the planet could become uninhabitable.[575]

Every time a politician denies that climate change is real, allows companies to pollute waterways, delays action on climate change mitigation, or leaves world climate agreements, we end up one step closer to *this* unrecognizable world.

Saturday, June 20, 2020, saw a record high temperature in Siberia, seventy miles north of the Arctic Circle. The temperature on this day was a whopping 100.4°F (38.0°C), the highest temperature ever observed in recorded history in the Arctic, a region that is warming more than twice as quickly as the rest of the world. Sunday, June 23, saw a temperature of 95.3°F (35.2°C) in the same spot, indicating that this temperature reading was not in error.[576]

Personally, I want this Earth to be habitable for my child and for my grandchildren, and beyond – but for that to happen, we cannot continue down the path we are on of environmental destruction and a lack of caring for the world around us.

To change this trajectory on a larger scale, we *must* vote every single time we have the opportunity. We must vote as if our lives depend on it (because they do, now more than ever). We must vote for leaders who put the environment ahead of environmentally destructive businesses, special interests, and lobbying groups such as the coal, oil, meat, dairy, and fishing industries. We must vote for leaders who advocate for, uphold, defend, and introduce new environmental protections, green technologies, and sustainable diets.

Besides the individual actions discussed throughout Part 2 of this book, voting is the most powerful tool we have in our arsenal, and is a way forward to avoid reaching tipping points where climate change will proceed unabated, species will forever become extinct, and we too will be in danger.[577,578]

If a safe and habitable world is what you want to see for your children, your grandchildren, and all future generations, use your power, use your vote, and vote as if your life depends on it. While you're at it, advocate for more walkable cities and walk or bike to your voting place.

Recipe 16 Walk, Bike, or Use Public or Shared-Ride Transit When Possible and Advocate for More Walkable Cities

As many of you are probably aware, the transportation sector is a significant source of greenhouse gas emissions – roughly 16 percent of total greenhouse gas emissions globally. The vast majority of these emissions come from carbon dioxide resulting from the combustion of petroleum-based products.[579,580,******] Methane and nitrous oxide are also emitted by the transportation sector, but in much smaller amounts.[581]

When talking about the transportation sector, we mean the movement of people and goods from point A to point B by cars, trucks, trains, ships, airplanes, and other vehicles. Of the total emissions that come from the transportation sector, the United States produces the most – roughly 3.7 percent, which is slightly more than the next two largest emitters combined – China (1.8 percent) and the EU (1.7 percent).[579,††††††] Of these emissions, 75 percent comes from road vehicles,

---

****** As mentioned in other sections of this book, greenhouse gas emissions from agriculture are at least 150 percent if not 300 percent greater than greenhouse gas emissions from transport (which is why diet and agriculture are so important in the fight against climate change).

†††††† If you are interested in finding more information about greenhouse gas emissions (from all sectors) around the world, spend some time exploring the following comprehensive and

including passenger cars and freight trucks, 12 percent comes from commercial aircraft, 11 percent comes from ships and boats, and 2.5 percent comes from rail and trains.[580,582,583]

Since 1990, greenhouse gas emissions from all sectors have increased by 141 percent, yet greenhouse gas emissions from the transportation sector alone have increased by 171 percent during the same time period.[581,584,585]

While this upward trend seems impossible to break – especially with a growing world population – we *can* reduce our emissions, and we *have* seen it be possible!

For example, during the height of the COVID-19 lockdowns in 2020, there were significant but temporary drops in greenhouse gas emissions globally.[586,587] During this short period, emissions decreased by 17 percent. Aviation saw a 75 percent decrease in emissions, and surface transport (primarily automobiles) saw a 50 percent decrease in emissions during the lockdowns; however, since transportation only accounts for 16 percent of total global emissions, the overall decrease was not as large as it sounds like it should be based on the significant decrease in the use of transport.[588,589,590,‡‡‡‡‡‡]

The lockdowns resulted in the most dramatic decrease in carbon dioxide emissions *ever* recorded.[585,590] Yet this decrease did not last long enough nor was it sufficient enough to cause meaningful progress toward meeting the goals set out by the Paris Climate Agreement.[588]

This means that now, more than ever, we – as individuals – must strive to lower our own personal emissions due to transportation (while also committing to engage in the processes outlined in some of the other important recipes addressed earlier).

Here are some examples of cities from around the world that have encouraged a decrease in the use of personal automobiles by facilitating the use of unmotorized transport such as walking and cycling. During the COVID-19 pandemic, Milan, Italy, turned 35 km (22 miles) of its streets into cycling and pedestrian ways.[591] Beijing, China, saw the use of bike-sharing systems increase by 150 percent, while Dundee, Scotland, saw bicycle traffic increase by 94 percent. New

---

interactive pinwheel chart from the World Resources Institute: www.wri.org/upload/circlechart2019/circle_state.htm.
‡‡‡‡‡‡ However, this is now seen as a "blip" in emissions, not an overall decrease, and sadly this has made essentially no dent in the overall upward trend in total greenhouse gas emissions.

York City saw a 67 percent increase in the use of bike-sharing systems, while in Philadelphia the use of bike trails increased by 151 percent.[592]

To accommodate the increase in foot or cycle traffic during the COVID-19 pandemic, some places installed bike lanes or closed streets to cars. Berlin, Budapest, Mexico City, New York, Dublin, and Bogota created pop-up bike lanes, and the governments of New Zealand and Scotland funded temporary bike lanes.[592]

The government of France transformed some car lanes into bike lanes in and around Paris to facilitate the use of alternative transport for commuters, and Brussels, London, and Seattle closed down streets to car traffic to allow more space for walking and bicycling.[593]

The closure of roads for foot traffic is a phenomenon that is being seen around the world, and one that may not have happened without the impetus arising from the COVID-19 pandemic. Some cities have even declared a more permanent closure of streets to automobiles in an effort to increase their bicycling capacity and walkability.[591,592]

This is absolutely a positive outcome stemming from the pandemic. The pandemic provided a unique opportunity to advance some green policy goals to transform some cities around the world to be more environmentally friendly and sustainable. In order to turn these changes into more permanent fixtures of environmental policy and infrastructure, it is more important than ever to vote for and elect government leaders who care about the environment and climate change.

It is imperative that these types of policies continue – and expand – after the pandemic. If we can make the most of this opportunity and make it easier for people to walk, bike, and commute safely without the use of fossil fuels, then we *can* make a more significant dent in our greenhouse gas emissions (from the transportation sector) that is more permanent, despite a growing global population and a projected increase in greenhouse gas emissions.

By solving one environmental crisis, we may have more time and resources to focus our efforts on some of the other areas discussed throughout this book where we can have just as much or even more of an impact on the environment. Remember, too, that walking or biking is good exercise for you – so this is a win–win!

So how do we put this into practice and actually lower our own personal emissions from transportation?

First, we can look at our current behaviors and identify areas where we can adjust.

Take a moment to think about how often you drive to the local grocery store, corner market, or pharmacy and ask yourself the following questions:

- How far is each of those car trips?
- How long do you sit in traffic to get there?
- Could you have walked there instead?
- Did you change your habits during the pandemic?
- Are there any "good habits" you can continue after the pandemic is over?
- Did you walk to stores more often during the pandemic so that you could get outside and breathe some fresh air?
- Did you plan ahead, make fewer trips, and ensure your purchases would last you longer during the pandemic to decrease your potential exposure to COVID-19?

If you *did* make some positive pandemic-related (emissions-reducing) changes, consider continuing them and making them a part of your everyday life. For example, continue to plan ahead for your grocery trips with a meal plan and a food list (such as those provided in Recipes 1 and 3) and make longer-lasting purchases (using the canning or jarring techniques described in Recipe 6). Consider shopping at farmers markets and try traveling shorter distances – remember, if the produce is local, it has not been transported very far, which can help you reduce your food's carbon footprint as well. Finally, walk or bike to your destinations wherever possible!

Taking these actions benefits your own health as well as that of the planet.

Obesity is at an all-time high around the world. Globally, 39 percent of adults over 18 are overweight and more than 13 percent are obese. Globally, 38 million children under the age of 5 are overweight or obese, and over 340 million children and adolescents between the ages of 5 and 19 are overweight or obese.[22] In the United States and other Western countries, between one-quarter to one-third (and sometimes more) of men and women are obese, and between half and two-thirds of adults are overweight.[594,595,596] Driving less and walking or biking more, in addition to eating a whole-foods, plant-based diet, are ways to be healthier and fight against health and environmental problems at the same time.

Furthermore, evidence suggests that time spent in nature and the outdoors has cognitive and psychological benefits for children, including improving their cognitive development, increasing their attentiveness, and reducing their exposures to air pollution.[597,598]

So instead of a long road trip, consider more local adventures on foot or bicycle. Instead of driving 40 miles (64 km) each way to and from the office, find out if you can work from home or telecommute. Instead of flying thousands of miles for a conference, see if you can attend it virtually, meaning more individuals can participate, save money, and protect the environment all at the same time.

By making some, if not all, of these changes, we can lower our own transportation footprint and reduce our risk of chronic disease simultaneously.

This pandemic has shown us that in-person meetings are not really necessary in many cases. And, if nothing else, this pandemic has shown us that we *can* reduce our impacts on the environment – there just has to be the will to do so.

When the will is there, so is the way. As individuals, we can continue many of the "environmentally friendly behaviors" that we started during the pandemic, even when the world returns to "normal."

Recipe 17 Use Less Electricity (and Energy) and Use Sustainable Sources Where Possible

> Our greatest weakness lies in giving up. The most certain way to succeed is always to try just one more time.
> —Thomas Edison

Most of the electricity generated around the world today still comes from the burning of fossil fuels, producing environmentally hazardous greenhouse gases.[599,600,601] Global electricity use is growing as a result of population growth and urbanization. Unless more of our electricity gets produced from renewable sources in the coming years, including wind, water, solar, and geothermal, greenhouse gases from this sector will only continue to rise.[602]

Here is a bit of information on each type of renewable energy source:

- *Solar energy* uses light energy – photons – that are absorbed by photovoltaic cells, known as solar panels. Photovoltaic cells convert photon energy into electricity – either direct current (DC) or alternating current (AC) – that can be used immediately or stored in batteries to be used later. Residential solar panels are at least 50 percent less expensive to install now (in 2020) than they were in 2011, making them more cost efficient than ever.

There are also solar thermal systems that can be used to heat water.[602]

- *Wind energy* is generated from wind turbines that convert kinetic energy from wind pressure into electricity – either for immediate use or to be stored in batteries. Currently, around 490 gigawatts of electricity are produced by wind power each year, preventing 637 million tonnes (702 million tons) of carbon dioxide from being released into the atmosphere.[603]

- *Wave and tidal stream energy* are generated from large underwater turbines. Similar to wind energy, spinning underwater turbines convert kinetic energy from water flows into electricity that we can use. Turbines may be part of a dam or they may be independent.[602]

    It is important to note that dams may have environmental repercussions. One such repercussion is that dams prevent mineral-rich silt from reaching river deltas, incidentally harming the dam or the reservoir itself. Another repercussion from dams is that they may block or prevent certain fish species from returning to their spawning grounds.[604,605,§§§§§]

- *Geothermal energy* is derived from the naturally occurring heat (hot water and steam) within Earth. Geothermal energy can be used for heating or cooling or can be harnessed to generate clean electricity.[606] Geothermal energy has been used in industry, aquaculture, and fruit and vegetable greenhouses for heating, bathing, and in other applications. It is one of the oldest methods used to generate electricity, and it produces more than 90 percent of the heating and electricity for Iceland.[606,607]

    Other countries, such as El Salvador, New Zealand, Kenya, and the Philippines, also use significant amounts of geothermal energy. Hawaii in the United States produces up to 24 percent of its electricity from geothermal energy. Producing electricity from geothermal energy is most efficient in places where there is a significant amount of tectonic activity.[606,607]

---

§§§§§ One notable case study in this area is the Snake River Dam in the northwestern United States. This dam has had numerous environmental consequences for the all-important Chinook salmon populations that were once in abundance in this area, and subsequently on the nutritional status and survivability of the Southern Resident orca population (previously discussed in this book).

Currently, too few countries get their electricity from renewable energy. But the number of countries that use renewable energy and the overall proportion of renewable energy use is growing.

As of 2020, the countries that use the most renewable energy are Iceland at 90 percent renewable energy, Germany at 13 percent, the UK at 12 percent, Sweden at 11 percent, Spain at 10 percent, Italy at 9 percent, Brazil at 7 percent, Japan at 5 percent, Turkey at 5 percent, Australia at 5 percent, and the United States at 4 percent.[602] Clearly, there is still a lot of room to grow in this area.

While most of us cannot personally dig a well and produce our own geothermal energy or put a windmill up in our backyards to generate electricity from wind power, we *may* be able to obtain some of our electricity from photovoltaic (solar-powered) energy. We *can* also purchase products that require less energy to run, and we *can* make changes within our homes to make them more energy efficient in order to reduce our personal energy use and to reduce greenhouse gas emissions from the energy sector (and at the same time save ourselves some money). We may also be able to engage in community-level renewable energy projects or vote for political leaders who support and engage with companies that *do* use renewable energy sources.

Here are some examples of how I have personally taken action to lower my energy use (and lower my household's energy bills by between 25 and 50 percent, depending on the season):

(1) *Insulate your windows:*[608] Having properly insulated windows helps keep cool air in and hot air out in the hot summer months and keeps warm air in and cold air out in the colder months. Insulating your windows cuts energy use from both heating and cooling, reducing costs significantly.

(2) *Use a solar heat-reflecting film or get "Low-E" glass:* We live in Southern California, where the weather is often quite lovely, at 70–80°F (21–27°C) during the daytime and the upper 50s to lower 60s°F (13–18°C) at night. With this weather, we are fortunate enough that we can keep our windows open most of the time for fresh air. However, there are rare days where the temperature can reach near 100°F (near 38°C) and we have to close all of the windows, which we put heat-reflecting film on so that our apartment stays at a comfortable temperature, without having to turn on the air-conditioning. If you live in a warm climate, consider

getting a "Low-E" glass or a window film that blocks up to 80 percent of solar heat while still allowing visible light to come through. If you do this yourself, as we did, you could save hundreds of dollars in cooling/air-conditioning bills.

(3) *Pack your freezer tightly:* The more tightly you pack your freezer, the more efficient it is (less energy is needed) and the better it keeps the freezer and refrigerator cold. Moreover, should you have a power outage, a full freezer will keep its contents frozen for far longer than a nearly empty freezer.[609]

(4) *Use compact fluorescent light bulbs (CFLs) or, even better, use LED bulbs to save significantly on electricity use:* More efficient indoor lighting is an important contributor to energy savings in most industrialized countries, as lighting uses approximately 20 percent of the total energy consumed worldwide (this figure is 18 percent in the EU and 30 percent in the United States).[610]

Classic incandescent and halogen light bulbs – the ones I grew up with – are very inefficient. By that I mean of the energy they consume, only 10–15 percent gets emitted as light – the rest is emitted as infrared heat.[610]

CFLs, on the other hand, produce significantly more visible light for the amount of energy they consume, making them far more efficient than incandescent lights.[610] CFLs use 75 percent less energy than incandescent bulbs and they last 6–15 times longer – clearly a better option than incandescent lights, which are fortunately becoming things of the past and are increasingly difficult to find.[611]

LED bulbs have taken several decades to evolve to where they are today as relatively common light fixtures. LEDs use 85 percent less energy than incandescent lights and have up to 50 times the life span.[610] This is even better than CFLs! These newer technologies are available on the market and help to improve the efficiency and reduce the daily costs of running our homes.

(5) If you do not live in a nice climate where you can open your windows, **adjusting your thermostat** is another way to use less energy. Adjusting your thermostat to a warmer temperature in the heat and to a cooler temperature in the cold can save 5–15 percent per year in heating and cooling energy use.[612] In the United States, nearly half of all the energy used in a home is used on heating and cooling.[613] In the EU, energy spent on heating and hot water is as much as 79 percent of energy use.[614]

(6) Finally, *turning off power strips connected to low-use electronics* may reduce your use of electricity as well. There are many so-called phantom or vampire devices that use power even when they are in standby or sleep mode. By unplugging them or fully turning off their power supply, you ensure that they do not use additional energy, saving you money and lowering your energy footprint.[608]

We can all be more mindful about our electricity and energy use. We can all take actions to use less electricity or make more environmentally friendly choices that will serve us better and at lower cost in the long run. We just have to *actively make these choices*, just as we actively choose what to eat on a daily basis or what type of transportation we will use.

By making more environmentally friendly choices, we do ourselves and the planet a favor. Many of these choices we can start making today.

Recipe 18 Educate Others! See, Do, Teach, and Lead by Example

This idea is near and dear to my heart, and it is an idea that I do every single day, whether with patients of mine in the hospital, in the graduate course I teach, in print or news media interviews, or in my own home.

See, do, teach.

As a registered dietitian at a major medical center, I see every patient as an opportunity to discuss healthier dietary patterns, whether for personal health or for the environment. On a daily basis, I meet people from every walk of life who are in hospital for a variety of reasons.

Some of my patients are in hospital because they have uncontrolled diabetes. Some of them are in hospital because they have had heart attacks. Some are in hospital for stomach or intestinal cancers and have had parts of their stomachs or intestines removed, making eating a challenging event every day. Some of my patients are even in hospital waiting to have one of their vital organs transplanted – it could be one of their kidneys, their heart, their lungs, or their liver.

Each one of these patients presents an opportunity to teach them how to eat more healthfully and on a very personal level. Foods are very personal to people. What we eat, why we eat, when we eat, and how we eat can play important roles in our physical, psychological, emotional, and social well-being.

Despite all of these various disease states being quite distinct from one another, the one thing they pretty much all have in common is that a plant-based diet is beneficial.*******

For example, when it comes to patients who have chronic kidney disease, eating more plant-based proteins instead of animal-based proteins helps the kidneys recover.[615] For patients who have heart disease that is related to diet (and most heart disease is), switching to a plant-based diet lowers their future risk of heart disease, decreases their cholesterol levels, helps them lose weight, and adds anti-inflammatory foods into their diet.[616]

For many patients who have cancer, whether of the breast, the stomach or intestine, or the throat and mouth, eating a plant-based diet or taking a plant-based nutritional supplement (including the use of plant-based tube feeds as needed) is more likely to help them recover and heal than eating animal-based foods.[617]

And for those patients of mine who are waiting to have one of their organs transplanted, I encourage them to eat a plant-based diet as it can help them stay healthier in preparation for the transplant.[618]

When I teach my graduate-level course "Nutrition and Chronic Disease" at the UCLA Fielding School of Public Health, I provide my students with a significant amount of literature to review that addresses the health and environmental benefits of a plant-based diet. Throughout the course, we cover topics including preventing and reversing chronic diseases through diet and other healthy behaviors. We discuss at length which countries are undergoing a nutrition transition and *why* they are doing so.††††††† We also discuss the environmental aspects of the dietary and nutritional intake patterns that I have addressed throughout this book.

Each time I teach this course, student feedback includes, "I was not aware of the extent to which plant-based diets are good for human health and the environment," and, "I learned valuable information that I plan to teach to others."

******* Please note that this does not substitute for medical or dietary advice. Please consult with your own doctor or dietitian before making any major changes to your diet, as there are chronic conditions/diseases that require specific and close monitoring of food choices and blood lab levels.
††††††† "Nutrition transition" means that a country as a whole is consuming significantly more animal-based food products and significantly more processed food products, including sugar-sweetened beverages and fast foods, as compared to their more traditional plant-based, unprocessed dietary staples.

Both of these comments are important for different reasons. First, if you are not aware of a problem, how can you address it? And second, once you are aware of a problem, what can you do to inform others about it?

Throughout my professional and personal lives, my goal has been to educate others about how the foods we eat either benefit our own health and the environment or are detrimental to both. This book is an extension of that goal to educate and promulgate that knowledge.

In my personal life, I am fortunate to be able to live a sustainable lifestyle. In a city where pretty much *everyone* drives a minimum of thirty minutes each way to and from work, I live close enough to my office that I walk to and from it every day. I am also fortunate enough to live within walking distance of my son's elementary school, which prepandemic he rode his bike to and from every day. We not only get the benefit of a close school, but we also get additional physical activity going there and back every day (as a round trip, it's roughly 1.5 miles or 2.4 km). I am also fortunate to live close enough to farms that the produce I buy at farmers markets and in supermarkets is locally grown, often organic, and relatively inexpensive because it did not have to travel far. Lastly, I am fortunate enough to have a supportive family that is interested in learning from me and is open to trying new things.

As an example, when I met my husband, he lived in Colorado and typically hunted elk every winter. As usually happens with newly formed couples, I – who had been mostly on a plant-based diet due to taking a vegetarian nutrition course in college back in the early 2000s – started eating more like he did (adding some animal products back into my diet) and he started eating more like I did (adding significantly more plant-based foods to his diet). We married in 2009, and soon afterwards, I started my doctoral studies at UCLA, which were focused on climate change, dietary patterns, and food security.

When I initially decided to eat a plant-based diet in the early 2000s, it was for health reasons: I wanted to avoid all chronic diseases. But after starting my doctoral studies, I learned so much new, underexposed, and undiscussed information about how dietary patterns affect environmental resources and contribute to environmental degradation – everything I talked about in Part 1 of this book.

Upon learning of these issues, I decided to resume my plant-based lifestyle, and I realized: *you can have it both ways.* You can eat for health and you can eat for the environment.

While my husband was initially skeptical, he followed a plant-based diet alongside me, perhaps begrudgingly at first! However, after just a few short months, he lost his taste for animal products, and now he purposefully chooses plant-based foods when we're eating out. He also enjoys looking out for new recipes for delicious plant-based foods.††††††††

We are now several years into our plant-based journey and he has not looked back. He no longer hunts animals, he prefers plant-based foods, and he even helps to educate his friends and family from time to time.

My own parents have opened their eyes, ears, and mouths to more plant-based food choices; and even my own son, in his short young life, seems to understand the benefits of a plant-based diet for a myriad of reasons – whether for health, for the environment, or for the animals themselves.

At the grocery store, other customers will frequently look in our cart and mention how healthy it is. When I mention I am a dietitian, it of course opens the door to many questions about diet and health. Whenever that happens, I nearly always take the time to explain *why* I eat a plant-based diet. While it is often a very abbreviated discussion, it is nonetheless a discussion, and if I can move the needle toward being healthier and more sustainable with even one person, that's a good thing!

This is also something I do every time I am interviewed for a news story, magazine article, or on the radio about nutrition. I address the personal health impacts of the foods we eat and point out the environmental effects of the foods we eat. I do this because not everyone knows about this. I do this because it is important to me to *lead by example and spread the word.*

Or, put another way, I "Fight for the things [I] care about, but do it in a way that will lead others to join [me]."§§§§§§§

All of the things I discuss in this book are actions we can take in our daily lives to feel more empowered about the world around us – to feel that we are doing something with purpose and that is useful and helpful for protecting the planet and our health.

Many individuals who were in lockdown felt as though they were doing nothing, but in truth most of us were putting the community

---

†††††††† He's the real cook in the house.   §§§§§§§ Ruth Bader Ginsburg.

and the health of others ahead of just ourselves. Helping others by teaching them about plant-based diets similarly helps protect the whole Earth.

We have no other home – we have nowhere else to go – so we must put the health of Earth and its habitability above the singular desire some feel for food products that harm the environment, whether they are animal-based products or foods made with palm oil.

That is what I encourage every one of you reading this book to do. Reach out to others who may be struggling with health issues, reach out to others who may be struggling with environmental fears, and discuss these issues with them. Discuss with them what you have learned and what you now know. Educate them and give them resources they can use to learn from as well. Lend them this book to read, or get them their own copy!

In Recipe 21, I provide a list of books and films that offer further information on some of the topics I have covered in this book: healthy diets, animal welfare, plastic pollution, environmental concerns, and more.

"Be[ing] the change you want to see" in this world starts with you, and it starts with me, and it starts with each and every one of us doing our part to make a measurable difference and to work with others, encouraging them to make a difference as well.

After you read this book, I hope that you will know more about these issues than you did before you read it. I also hope that you will have more tools in your tool belt to make you better equipped to live a more sustainable lifestyle, eat a more sustainable diet, speak up more about sustainable living and eating patterns, vote for and support these important issues, and lead by example. The time to do these things is now.

Recipe 19 Plant Trees and Other Plants to Protect the Environment

> He that plants trees loves others besides himself.
> —Dr. Thomas Fuller

Around the world, forests are being razed and burned down at ever-increasing rates – over forty football fields-worth of trees and other plants every minute. During the first four months of 2020, roughly 1,200 square kilometers (464 square miles) of Amazon rain forest were cut down, which is an increase of almost 55 percent compared with how much rain forest was cut down during the first four months of

2019.[619,********] Most of this deforestation is being done by people who are clearing the land for the production of soy, cattle, palm oil, or wood. This is often done illegally or under the radar and may be fueled by local corruption, politics, and/or conflict.[620]

Brazil currently has the highest rate of forest loss, and the Democratic Republic of the Congo (DRC) has the second highest rate. Colombia has also seen an increase in deforestation as a result of the disarming of its armed forces, which has paved the way for small, armed groups to illegally clear the land for cocoa farming, mining, and logging.[620]

The rate at which we are losing trees and other forested areas is disastrous for the wildlife that depends on them and for the indigenous peoples who live within them. Losing rain forests also significantly hinders the fight against climate change, as perhaps even more import-ant than their role in creating oxygen, forests absorb and store huge amounts of carbon. In fact, forest conservation and replanting could do as much as 30 percent of the work that is needed to meet the goals set by the Paris Climate Agreement: to lower the total amount of carbon in the atmosphere so as to avoid global warming of greater than 2 °C.[620,††††††††]

Tragically, during the COVID-19 lockdowns, illegal mining and deforestation increased, as did the deregulation of environmental policies. Moreover, many governments were immersed in fighting the pandemic and so lacked the funds and the attention to work on conser-vation.[619,621] For a more in-depth discussion on these topics, please return to Part 1 of this book.

So What Can We Do?

Anything *we* can do to combat deforestation by planting more trees or other plants is imperative. Trees and other plants are vital for a myriad of reasons, including:

(1) Removing carbon dioxide from the atmosphere, absorbing it, util-izing it, storing it, and replacing it with all-important, life-giving oxygen;

******** This is nearly the equivalent of 225,000 football fields per year; approximately 1,860 football fields per day or 77 per hour (~1.3 per minute). This is a conservative estimate.
†††††††† As mentioned in previous recipes, regenerative farming could advance carbon sequestration even further and lower carbon in the atmosphere even more.

(2) Transforming carbon from the atmosphere – through photosynthesis – into chemical energy and carbohydrates, namely glucose and fructose, which are regularly consumed by animals, including us, for energy;[622]

(3) Significantly improving the productivity of habitats and ecosystems by altering the structures, distribution, and functions of those ecosystems.

Forests, trees, and other plants also improve food security by maintaining the environmental conditions that are needed for agricultural production, including stabilizing the soil, preventing erosion and salinization, increasing soil fertility and carbon levels, enhancing the land's capacity to store water, and moderating the impacts on air and soil temperatures from environmental stressors.[623,624,‡‡‡‡‡‡‡‡]

As one example, where coffee trees are grown in forested areas, called "shade-grown coffee," the final product is often a higher-quality coffee bean with a higher yield that does not require the use of fertilizer. Furthermore, shade-grown coffee plantations protect important habitats for birds who eat insects and other pests that typically harm coffee beans. It is suggested that one single bird can save 23–65 pounds (10–30 kg) of coffee per hectare of forest from pests every year while reducing or removing the need for expensive and dangerous chemical pesticides.[625]

Trees are also important because they affect local microclimates by regulating local temperatures, humidity levels, moisture availability, and light conditions. The success of many agroforestry systems around the world depends on the ability of trees to moderate soil and air temperatures and to increase relative humidity levels within the system.[367,623,§§§§§§§§]

Larger-scale forests affect the global climate through their influence on rainfall patterns, surface reflectivity, and certain other

---

‡‡‡‡‡‡‡‡ Salinization is the accumulation of water-soluble salts in soil. This is often detrimental to crops, hindering their growth and limiting their ability to take up and assimilate water.
§§§§§§§§ This is also known as the "small" water cycle, which is land-to-land precipitation–evaporation–precipitation, and it accounts for 50 percent of the precipitation that falls on land. The precipitation of the small water cycle is characterized by light, frequent rainfall, mist, dew, and short showers – the kind of precipitation that builds vegetation. This is in contrast to the "large" water cycle, in which water evaporates from the oceans, forms clouds, and falls on land (which accounts for the other 50 percent of precipitation that falls on land). Disruptions to the small water cycle are more deleterious to the health of the land and soil than are disruptions to the large water cycle (Tickell, 2017).

meteorological variables.[623] When forests are destroyed, the way sunlight reflects off Earth's surface changes. In a forest, sunlight is absorbed by leaves, branches, and tree trunks, among other things, cooling the air temperature. Without forests, sunlight is reflected, trapping more solar heat and causing daytime temperatures to increase.[623]

This "heat-island effect" is a phenomenon seen in urban environments and cities, which are known for being significantly warmer than their "greener" surrounding areas. In fact, temperature differences of more than 9°F (5°C) have been observed between city centers and more vegetated suburban areas.

Urbanization is associated with a 1°F (0.56°C) increase in average temperature every decade. This warming is due to paving, the number of buildings, the electricity demands, and the lack of vegetation, and is further exacerbated by climate change and global warming.[626]

Urban forests and tree cover can ameliorate some of this heat through shading, which reduces the amount of heat energy absorbed and radiated by built surfaces. Trees can also cool down their surroundings through evapotranspiration.********* Additionally, trees can modify air flow, affecting the transport and diffusion of heat, water vapor, and pollutants.[626]

Shade trees on housing properties have been shown to substantially lower the cost of cooling a house. Shade trees also benefit a community through their aggregate effect on the urban climate.[623] Other evidence suggests that shade trees rejuvenate cities by creating healthier environments, positive community interactions, and new jobs such as tree-planting crews or foresters.[626,627]

Moreover, there is a strong bond between people and the natural environment, one that is often associated with increased mental acuity, calmness, and emotional well-being. Perhaps these effects are related to our evolutionary past – after all, we did develop within forested environments. It may also be that these effects are related to the feelings of relaxation that seem to occur when we see trees and other vegetation, which lower blood pressure, slow down the heartbeat, and improve brain wave patterns.[628]

********* Evapotranspiration is the process by which water is transferred from the land to the atmosphere by evaporation from the soil and other surfaces and by the transpiration from plants.

A number of experimental psychology studies link exposure to nature with increased energy, a heightened sense of well-being, feeling more alive, increased feelings of happiness and health, and decreased feelings of exhaustion.[629] Perhaps this is due to the idea of "being one with nature," or perhaps it is the extra oxygen surrounding trees and other plants that provides some of these purported benefits.

One large tree provides enough oxygen for up to four people each day and absorbs more than forty-eight pounds of carbon dioxide from the atmosphere. It is estimated that one acre of mature trees can provide enough oxygen for eighteen people per year.[630] With 7.8 billion people on the planet, and at least 2–3 billion more expected, we will need more trees and plant cover, not less.[631]

While phytoplankton and microalgae in the oceans are responsible for producing more than half of the oxygen that we breathe, rain forests produce 28 percent.[632] We cannot plant phytoplankton in our neighborhoods or our cities, but we *can* plant trees, which is why, although they provide less oxygen than we get from the ocean, it is necessary to discuss them, because planting trees is something we can *actually do*.[633]

Trees, rain forests, and other plants enhance and support biological diversity, stabilize and support various microclimates, and provide a number of ecosystem services.[634] For many indigenous populations around the world, trees, forests, and rain forests in particular are their lifelines, their livelihoods, their medicines, their shelters, and their foods.[635]

For the rest of us, who live far away from the natural world, in cities, in high-rises or walk-ups, or in single-family homes with a yard, having a bit of nature close to us can provide some respite from the day-to-day frenzy many of us experience.[636]

I can tell you that having trees in my neighborhood helped my well-being immensely during the COVID-19 lockdowns. Being able to look at tree tops while on my balcony, seeing the wind blow through them, and watching birds and squirrels flitter around their branches brought a fresh dose of mental clarity every day when going outside was strongly discouraged.

Similarly, counting the various tree species in our neighborhood was a wonderful diversion for my too-often cooped-up son. So the next time someone tells you to "stop and smell the roses," perhaps you can offer up a "stop and admire the trees."

Recipe 20 Seek Out Cotton or Other Natural Fibers, Not Synthetics

While clothing *can* be made from natural substances such as cotton, flaxseed (linen), or hemp, many items that we wear today are actually made from a blend of synthetic fibers derived from coal, petroleum, and natural gas.[637]

Rayon, nylon, polyester, acrylic, and spandex are just a few examples of the synthetic fibers that many of us wear. The following provides some more in-depth explanations on how they are produced and what their environmental impacts are.[637]

Rayon is known as artificial silk since it resembles silk in its appearance. Rayon is often used to make shirts, ties, bed sheets, and bandages. It was invented over 150 years ago and became widely available in the 1920s. To produce rayon, tough plant materials (such as the cellulose derived from wood pulp) are chemically and mechanically processed using sodium hydroxide and carbon disulfide, and they are then spun into threads using sulfuric acid.[638]

The chemicals used in rayon factories are not environmentally friendly. Many of these chemicals end up in rivers, waterways, or drinking water and poison communities downstream. This has been seen in factories in China, Indonesia, and India, where workers sometimes wash chemicals off rayon textiles and dump chemical effluents from factories into rivers or waterways.[639] It is worth noting that these are not rogue factories, but rather are part of large brands that provide rayon fabric directly to major fashion brands.[638]

Reports also note that individuals who work in rayon-producing factories are at an increased risk of nerve damage, heart disease, and stroke as a result of their exposures to these harsh chemicals.[638]

Rayon can be made from bamboo, and it is often touted as an eco-friendly fabric, primarily because bamboo grows quickly and without the use of pesticides or herbicides. However, it is actually a misnomer to call this product "eco-friendly," since bamboo pulp, like its wood pulp counterparts, still needs to be treated heavily with toxic chemicals. Stores that sell bamboo-derived rayon clothing or fabric frequently label it simply as "bamboo," making it difficult to identify as rayon.[638]

Just like native forests in Indonesia are destroyed to produce palm oil, rain forests in Indonesia, Canada, and Brazil are being slashed and burned to grow plants that will be turned into cellulose-rich fabrics such as rayon. Unfortunately, though, only 30 percent of a tree's pulp is used to produce rayon. The rest is usually burned, producing greenhouse gases while simultaneously being eliminated as a carbon sink due to deforestation.[638] So although rayon is made from plants and sounds like it should be eco-friendly, it is not.

Nylon is another synthetic fiber that was first produced in the early 1930s by scientists at DuPont Company. Nylon's main ingredients are coal, water, and air.[637] Nylon has elastic properties and does not lose its strength even after repeated use. Nylon is also easy to wash and dry. Nylon is used most often in socks, stockings, tents, umbrellas, parachutes, tarpaulins, floss, and in toothbrush bristles. Nylon threads are used to make fishing nets, climbing ropes, and strings for tennis racquets.[637]

Nylon is one of the most widely used synthetic fibers, but it too is not environmentally friendly, polluting both the air and water.

The manufacturing of nylon produces nitrous oxide, a greenhouse gas that is around 280 times more potent than carbon dioxide in its warming abilities and is partially responsible for depleting the Earth's ozone layer.[32,640,641] Nylon does not biodegrade, which in the case of discarded fishing nets and other nylon products is extremely hazardous to ocean wildlife, as discussed throughout Chapter 4.

Polyester is an umbrella term used for a few different fibers. Polyesters are known for their strength, for being lightweight, for having good elasticity, for being resistant to wrinkling, and for being easy to wash and quick to dry. The most commonly used polyester is Terylene, which is often blended with natural fibers, such as cotton (called Terrycot), to improve the way it feels. Terrycot absorbs liquids better than Terylene alone does. When Terylene is blended with wool (called Terrywool), it provides warmth in addition to the other characteristics typically associated with polyester.[637]

Polyesters are used to make lightweight sails and are often used for making shirts, skirts, and suits. Polyesters can also be used to make Mylar, a film used in magnetic recording tapes, audio cassettes, video cassettes, and floppy disks (most of which are now obsolete), or it can be used to make conveyor belts.[637]

Unfortunately, polyesters require large amounts of water to produce (for cooling purposes), along with lubricants, which can contaminate waterways. Polyesters are the most widely used synthetic fibers, and to their benefit, can be recycled.[640,641] Additionally, polyesters *can be made* from recycled plastic bottles (which are made from polyethylene terephthalate), which could save over 2.4 billion bottles from going to landfills each year. Another benefit of using recycled materials is that doing so can reduce the amount of air pollution from the production of polyesters by up to 85 percent.

One purported disadvantage of recycled polyesters is that their quality is not thought to be as good as that of "virgin" polyester.[641] But it certainly seems that the benefits of creating recycled polyester could outweigh the issue of quality.

Acrylic is another human-made synthetic plastic material containing a derivative of acrylic acid. The most common acrylic plastic is polymethyl methacrylate (PMMA), commonly known as Plexiglas, Lucite, Perspex, and Crystallite. PMMA is a tough, transparent material that is resistant to ultraviolet radiation and weathering. It can be colored, molded, cut, and drilled, allowing it to be used in many applications, including in airplane windshields, skylights, automobile taillights, and aquariums. Acrylic is also used to make sweaters, socks, shawls, carpets, and blankets.[637,642]

To manufacture acrylics, highly toxic substances – which require careful storage, handling, and disposal – are used, and they may cause explosions if not properly stored. Many of these chemicals produce toxic fumes. This has led to legislation requiring acrylic manufacturing to be carried out in closed environments that can capture, clean, or neutralize these fumes before they are discharged into the atmosphere.[642] Acrylics cannot be easily recycled and often end up in landfills, where they do not biodegrade.

Spandex, also known as Lycra (or elastane), was invented in 1959 by a DuPont chemist. Spandex is a petroleum-based product that requires a lot of energy to produce.[643] Spandex is known for its elasticity and is often used in clothing that requires a snug fit, such as swimwear and leggings. Spandex is often mixed with other fibers such as cotton to produce fabrics that stretch.[637]

Although spandex is less widely used than polyester or nylon, it is estimated that 80 percent of all garments contain some spandex. Moreover, many materials traditionally made from 100-percent cotton, such as jeans, are now mixed with spandex to improve their comfort, mobility, and fit.[643] Unfortunately, however, like many other synthetics, spandex is not biodegradable and often ends up in landfills or the oceans.

While I was researching all of these different synthetic fibers, I came across a very interesting study that asked the following question: "What happens to clothing fibers when they are washed?"

Case Study

To answer this question, researchers laundered polyesters, polyester–cotton blends, and acrylic fabrics under a variety of conditions, including at various temperatures and with various detergent types and conditioners. This study found that, on average, 700,000 fibers get released from a load of acrylic fabrics weighing 6 kg (13.2 pounds). It also found that as many as 1 million fibers may get released from washing just a single polyester fleece.

It is worth noting that the number of fibers that get released varies by the type of fabric, the wash treatment, and the type of machine used. Acrylic clothing sheds the most fibers, followed by 100-percent polyester clothing. Meanwhile, polyester–cotton blends shed the fewest fibers (of those tested). Top-loading washing machines released seven times as many fibers as did front-loading machines.[644,645] The average size of the shed fibers ranged from 12.0 to 17.5 μm in diameter (~1/6 the diameter of a human hair) and from 5.0 to 7.8 mm in length.[646,647]

What this means is that every one of us may be releasing as much as 790 pounds (360 kg) of microscopic plastic fibers into the environment each year as a result of the clothes we wear and how we wash them. Some of these fibers, which are too small to be captured by washing machines, have been found in the effluent of sewage treatment plants, and they contribute to the microplastic pollution found in the oceans and other aquatic environments.[646,648,649]

As was mentioned throughout Chapter 4, microplastics, including fibers from clothing, often get consumed by sea life and eventually by us.[203,646] In one study, 73 percent of fish were found to have polyethylene, methyl cellulose, and nylon fibers, among others, in their gut contents.[650] These microplastics contain many persistent organic compounds and detrimental chemicals, so we need to consider what we wear as part of the overall picture with regards to plastic pollution and to possible solutions.

Wherever scientists look, plastic fibers contaminate the environment and the foods we eat or the beverages we drink, including seafood, beer, salt, and tap water.[651,652] Although plastic pollutants in the ocean come from multiple sources, as previously discussed, the International

Union for Conservation of Nature (IUCN) estimates that as much as 30 percent of oceanic microplastics come from synthetic textiles that are more commonly manufactured in developing nations, which do not have stringent environmental regulations or wastewater treatment facilities that can sufficiently filter them out.[653]

And while it is not likely or even possible that everyone in the world will go back to wearing 100-percent natural fibers, such as cotton, hemp, silk, or linen, having a better understanding of how clothing and other textiles are produced at least allows us to make better-informed and more eco-conscious decisions on what we buy and perhaps how we clean it.

Natural Fibers

Although nothing in this world comes without a price, natural fibers do *tend to be* more sustainable and eco-friendly than the synthetic fibers discussed above.

Cotton comes from a plant that is grown on 2.5 percent of the world's cultivated land. Conventional cotton, though more sustainable than synthetic fibers, uses nearly 16 percent of the world's insecticides and large amounts of nitrogen-based synthetic fertilizers, which are harmful to the environment (approximately one-third of a pound of fertilizer per pound of cotton).[††††††††] In case you didn't know, it takes roughly one pound of raw cotton to produce one t-shirt.[654]

There *are* organic cotton farms that do not use synthetic chemicals or genetically engineered seeds, making them better for the environment than both traditional cotton and synthetic fibers. It may take a little extra research on your part and it may represent a small increase in price, but it is possible to buy sustainable, organic cotton clothing.

In addition to organic cotton, *linen* (made from flax) and *hemp* are also natural, plant-sourced fibers that are sustainable and have been used in textiles and clothing.

Flax and hemp are plants that have been cultivated for thousands of years on almost every continent, and they have been used to make clothing, ropes, and sails. Linen and hemp can be blended with other natural fibers to create fabrics that are both durable and soft.[655] Some of the benefits attributed to these plants include that they do not

---

†††††††† Much fertilizer runs off into freshwater habitats or groundwater and contributes to the creation of the oxygen-free dead zones discussed in Chapter 3.

require harsh chemical herbicides and they naturally repel pests. Moreover, flax and hemp recycle up to 60–70 percent of the nutrients that they take in from the soil.[655] Flax and hemp have antimicrobial properties. They also act as carbon sinks, meaning they absorb carbon dioxide from the atmosphere, store it in the soil, and replace it with oxygen. Synthetic fabrics cannot do that.[656]

Other benefits of flax and hemp are that they require comparatively little water to grow and to process (roughly one-quarter of the amount of water that cotton requires).[657] Natural fibers are also recyclable; they can be refabricated repeatedly after end use and processed into new textiles or articles of clothing.[656] Furthermore, natural fibers such as cotton, silk, wool, hemp, linen, and bamboo are also biodegradable where synthetic fabrics are not.[658]

Silk and wool, while highly renewable, natural, recyclable, and compostable – all good things – do have potential ethical concerns associated with them, and it may require research on your part in order to find where you can obtain ethically sourced silk and wool.

Traditionally, *silk* is obtained from the cocoons that are spun by silk moths after they have metamorphosed and flown away. Silk moths feed on mulberry leaves (which do not require pesticides or fertilizers), making silk (the fiber itself) relatively safe and innocuous. What makes silk ethically dubious, however, is that large silk companies today use domesticated silkworms that live only a few days and then die soon after laying their eggs and producing their silk cocoons.[659],[‡‡‡‡‡‡‡‡‡]

There are still some companies that produce silk made from the cocoons of wild and semiwild silk moths. This is known as ahimsa silk and does not require the killing of the silkworms. Unfortunately, however, finding these companies can be tricky.

Wool may also have ethical concerns, primarily associated with the wool-shearing process and the enteric methane emissions of the sheep themselves.

Sheep are prey animals that are prone to panic when held down, which they undergo during the shearing process. Moreover, shearing may be a painful or potentially terrifying ordeal for the sheep, as the shearers are often paid by the "volume" of wool they obtain as opposed to by the

---

[‡‡‡‡‡‡‡‡‡] The silk cocoons are boiled with the pupae inside (this kills the pupae prior to the completion of their metamorphosis), meaning that these domesticated silkworms do not live out their natural life spans.

hour. This can lead to quick work that may be less sensitive to the animal's welfare and has resulted in abuse, mutilation, and skinning.[660]

The other factor mentioned is that sheep release huge amounts of methane – as much as 30 liters each day (cows can release up to 200 liters per day).[661] Their manure also emits significant amounts of greenhouse gases.

It is worth noting that the primary environmental boon for wool is that it can be reused, recycled, and repurposed over and over again, which lowers its overall environmental footprint and lessens the need for new wool.[662] Animal welfare is a concern with wool, but this can certainly be addressed, helping to turn wool into a more environmentally friendly and ethical choice.

This all matters because 80 billion pieces of clothing are purchased every year, but only 10 percent of those purchases end up donated, recycled, reused, or repurposed. The other 90 percent of clothing that we buy ends up in landfills (eventually), where they emit methane gas or may release some of their chemicals into the environment.[658] It is no wonder that fashion is among the top three industries contributing to climate change and environmental degradation (transport/oil and meat production are the other two).[658]

This is yet another indication that what we do matters – that what we purchase and our choices matter. With more knowledge, we can become more thoughtful consumers and make choices that help the environment, whether that is choosing more environmentally friendly fabrics or being more mindful about how we discard, recycle, or repurpose items we no longer use.

As a helpful summary, Table 8.11 may assist you in making some of these choices.

Recipe 21 Read These Books and Watch These Films

> Education is the most powerful weapon which you can use to change the world.
> —Nelson Mandela

As someone who enjoys reading and viewing documentaries on the subjects I have covered throughout this book, I thought it might be useful to provide a list of books and films that go into more depth (and with more imagery) on some of these topics (Table 8.12). I personally

Table 8.11 Various fabrics and their environmental impacts

Textile	Natural/ synthetic?	Biodegradable?	Reusable/ recyclable?	Pollutes the environment?
Rayon	Produced from wood pulp that comes from mature forests (natural, but not sustainable), also requires many harsh chemicals to produce	Yes, but often sent to landfill	Yes, but rayon fibers are not as of yet recyclable[641]	Yes, most often during processing
Nylon	Completely synthetic, made from petroleum sources	No	Yes, 100-percent nylon is recyclable, but subsequent products may not be as lasting	Yes, when fibers end up in marine environments
Polyester	Completely synthetic, made from petroleum sources	No	Yes, 100-percent polyester is recyclable, but subsequent products may not be as lasting	Yes, when fibers end up in marine environments
Acrylic	Completely synthetic, petrochemical feedstock	No	Not easily recyclable, can be transformed or downcycled into new objects[663]	Yes, toxic chemicals are required to manufacture; when used in clothing, acrylic-blend fibers end up in in wastewater effluent
Spandex	Completely synthetic,	No	100-percent spandex can be, but not easily,	Yes, primarily on the production

Table 8.11 cont'd

Textile	Natural/ synthetic?	Biodegradable?	Reusable/ recyclable?	Pollutes the environment?
	polyurethane plastic base		and blended items cannot be recycled or reused[664]	side. as it requires a lot of energy to produce
Cotton	Natural; however, many of the fertilizers, herbicides, dyes, and finishing chemicals used are not	Yes	Yes; however, if it is a cotton blended with synthetic materials, then no	Primarily on the production end; end products, if 100-percent cotton, do not
Linen	Natural, requires few herbicides or pesticides	Yes	Yes, the fibers can be reused or remanufactured	The fabric itself does not, but the dyes may
Hemp	Natural, requires few herbicides or pesticides	Yes	Yes, the fibers can be reused or remanufactured	The fabric itself does not, but the dyes may
Silk	Natural, requires no herbicides or pesticides, produced from the domesticated silkworm; may have some ethical quandaries	Yes	Yes	The fabric itself does not, but the dyes may
Wool	Natural, produced from sheep; may have some ethical quandaries	Yes	Yes	Yes, the methane from enteric fermentation in the sheep, their manure, and dyes

Table 8.12 Book and film recommendations

Book recommendations	Film/documentary recommendations
What Animals Think and Feel – Carl Safina	*The Ivory Game* – Richard Ladkani
The World Is Blue – Sylvia Earle	*Sea of Shadows* – Richard Ladkani
The End of the Line – Charles Clover	*The Cove* – Louie Psihoyos
Eating Animals – Jonathan Safran Foer	*Racing Extinction* – Louie Psihoyos
We Are the Weather – Jonathan Safran Foer	*The Game Changers* – Louie Psihoyos
Four Fish – Paul Greenberg	*Food Choices* – Michal Siewierski
Homo Deus – Yuval Noah Harari	*Naledi* – Geoffrey Luck and Ben Bowie
Voyage of the Turtle – Carl Safina	*Echo* – PBS documentary
Spillover – David Quammen	*Before the Flood* – Fisher Stevens and Leonardo DiCaprio
An African Love Story: Love, Life and Elephants – Daphne Sheldrick	*Takeout* – Michal Siewierski
Rattling the Cage – Steven Wise	*Elephant Queen* – Victoria Stone
The Elephant Whisperer – Lawrence Anthony	*Forks over Knives* – Lee Fulkerson
The Last Rhinos – Lawrence Anthony	*Chasing Coral* – Jeff Orlowski
Babylon's Ark – Lawrence Anthony	*Chasing Ice* – Jeff Orlowski
Upheaval – Jared Diamond	*An Inconvenient Truth* – Davis Guggenhein
Collapse – Jared Diamond	*An Inconvenient Sequel* – Jon Shenk
Sapiens – Yuval Noah Harari	*Blackfish* – Gabriela Cowperthwaite
Behind the Dolphin Smile – Richard O'Barry	*The End of the Line* – Charles Clover
The China Study – T. Colin Campbell	*Unlocking the Cage* – Chris Hegedus
Silent Spring – Rachel Carson	*A Plastic Ocean* – Craig Leeson
The Sixth Extinction – Elizabeth Kolbert	*Plastic Paradise* – Angela Sun

Table 8.12 cont'd

Book recommendations	Film/documentary recommendations
Kiss the Ground – Josh Tickell	*Kiss the Ground* – Josh Tickell
Tomatoland – Barry Estabrook	*David Attenborough: A Life on Our Planet* – Jonathan Hughes, Alastair Fothergill, and Keith Scholey

have read every one of these books or have watched every one of these films and can attest to their educational value.

Should you choose to read or watch beyond this list, please see the complete list of references provided at the end of this book. Also at the end of this book, I provide you with quick and easy "cheat sheets" or "recipe cards" for each idea I have discussed in this book – see the Appendix. Here you will find brief summaries or recipe cards of each idea that you can hang on a wall or turn to for a quick and easy reminder of some of the key ideas/takeaways from this book.

9 EPILOGUE/CONCLUSION
Final Thoughts and Key Takeaways

As a group, we humans produce a lot of greenhouse gases that affect this world in many ways – seen and unseen. Our greenhouse gases make it difficult for many species – who have lived on a mostly stable planet for millions of years – to adapt to new warmer, drier, and more erratic conditions. Their survival, whether they live in the Arctic, the Antarctic, or the rain forests of the tropics, depends on us gaining control of and reversing the harmful greenhouse gases we produce and restoring the balance of nature.

Our survival depends on this, too.

Increasingly, we are seeing temperature spikes in what are supposed to be some of the coldest places on the planet. We are seeing changing weather patterns and an increase in the frequency and severity of hurricanes, floods, droughts, and fires. The year 2020 saw thirty named tropical storms and hurricanes in the Atlantic region – a new record. A number of these storms also had some of the most rapid rates of intensification ever observed.

Around the world, the 2020 fire season set a number of new and grim records. In the United States, over 8.2 million acres of land (33,000 km²) were lost to fires in California, Oregon, and Washington states. Colorado saw its largest wildfire season ever, with 625,000 acres (2,530 km²) lost to infernos. Australia, too, saw its worst fire season in history, with over 44 million acres of land (180,000 km²) turned to ash and nearly 3 billion wild animals scorched, harmed, or killed.[665]

Over 800 million people around the world are malnourished or are at risk of malnutrition because they cannot get enough to eat.

The COVID-19 pandemic has made this worse for many. And yet nearly 2 billion people around the world are overweight or obese, consume far more than they need to for good health, and die younger than they should.

These are not problems of calorie production (total crops produced). We produce more than enough food to feed every person on Earth. The problem is that we do not distribute this food appropriately. Instead of feeding those who need it the most, we send vast quantities of calories to feed domesticated animals while simultaneously discarding huge amounts of food. Both of these result in wide-ranging harms to the environment.

Over the last few decades, the number of domesticated animals on this planet has tripled if not quadrupled, and the number of animals that are killed each year to feed the growing human appetite is staggering: over 50 billion chickens, 1.5 billion pigs, half a billion sheep, nearly 300 million cattle, and over 150 million tons of sea creatures.[666]

In contrast, the numbers of individuals of unique wild species – elephants, giraffes, rhinoceros, lions, tigers, polar bears, pangolins, vaquita porpoises, and so many others – have fallen to the point that their extinctions may be inevitable.

Is this the world we envision for ourselves, our children, our grandchildren, and beyond?

I certainly hope not. Instead, I hope that this book has given you a number of ways to make a positive difference to the world you live in. In fact, the key idea I would like to see you take away from this book is that you – as an individual – can make a difference. Your actions, your behaviors, and your choices *matter*. What you choose to do today, tomorrow, the next day, and so on affects Earth and the way it functions.

While each recipe provided in this book may *seem* like a large undertaking or even a daunting task, or you may feel as though you are just one person and making too small a change to really matter, it is important to recognize and acknowledge that any one change or alteration you make as a result of reading this book *can* in fact make a big difference to your health, to your wallet, and to the environment. So why not start today? Choose one recipe to get you going, to feel more empowered, to take action. Then choose another one, and then another one.

Know that the foods you choose to eat *can contribute* significantly to environmental rehabilitation (or degradation).

Know that as our human population grows your food choices will have more of an impact on the environment than any other activity you personally engage in on a daily basis.

Know that by choosing to eat a plant-based diet and significantly less (or, even better, no) meat, dairy, and fish products you are helping your own health and the environment.

It's really that simple. Your choices, your behaviors, whether or not you choose to eat a plant-based diet, whether or not you choose to recycle, whether or not you choose to drive or to walk to the market, and whether or not you choose ethically sourced products and services make a difference in the fight for environmental sustainability.

We absolutely cannot continue on the path we have been on and "hope for the best."

Sadly, many of us know that fossil fuels are a major source of greenhouse gases, and yet too many of us still drive petrol-guzzling vehicles.

Many of us know that plastic is bad for the environment, and yet too many of us still buy that single-use bottle of water and toss it in the trash.

Too many of us eat animal protein-heavy diets and then wonder why we are dying too young and getting heart disease, cancer, or other chronic conditions that take the life out of our days and the days out of our lives.

Before this book, perhaps too few of us knew that what we eat takes a toll on the environment. But not anymore.

We must each take personal responsibility to do our part, eat a more sustainable diet, buy and use fewer things, reuse what we can, and live in environmentally respectful ways.

We only have this one planet. We do not have another. As the saying goes, "There is no planet B." And while we may never restore Earth to what it was before the Industrial Revolution, we can work toward halting and perhaps even reversing some of the damage we have done.

We have seen that when parts of Earth are left alone to rebuild, they will do so.

We have seen that when parts of the ocean are left to their own devices, they will recover.

If we have the will to protect Earth by changing our dietary habits, our behaviors, our consumption, and our use *of* the planet and its species, we will probably find that Earth can recuperate and perhaps even thrive.

But there is no time to wait – there is no time to deflect. If you want less disease on this planet, fewer viruses jumping from animals to humans, fewer lockdowns, and less death, change must happen – starting now, with *your* help!

We must choose to be more sustainable. That's how this works. And that's what I have addressed throughout this book.

So start today by significantly reducing your intake of animal proteins and increasing your intake of plant-based foods.

Advocate for regenerative farming techniques.

Vote for lawmakers and government officials who support sustainable farming and environmental policies.

Drive less and walk or bike more.

Use less fossil fuel-based electricity and lean into renewable energy as if your life depended on it.

Refuse to use plastic, but if there is no other option, reduce your use, reuse, and recycle as much as possible.

Seek out sustainable clothing and eco-friendly food products that are fair trade and pay a living wage to those who produce it.

Be more conscious about your decisions and seek to be a responsible consumer.

Reduce your level of consumption to be more in line with preserving this one and only home that we have (and its Earthlings).

Do these things now – any one or more of them – to lower your impact on the planet, or run the risk that too many species will be lost to extinction, perhaps even our own species.

It does not have to be all or nothing. Do the best that you can do, be proud of your progress, and work toward taking each recipe one step further today than you did the day before. Each recipe in this book represents a journey. You do not have to do them all at once, and you do not have to be perfect. I know I'm not, but I try to do the best I can every day for myself, for my son, and for the planet.

Like having a child, this book has been a labor of love for me. I have spent more than five years researching it, writing it, and honing its message. I have learned more than I ever thought possible and have lived every aspect of this book to both model the behaviors and actions I wish to see in this world and to be able to effectively describe and prescribe what I consider to be some of the most effective actions we can take as individuals.

In fact, writing this book has almost been like writing a memoire for me.

Without knowledge, I cannot expect individuals such as you to be aware or to care. But with understanding, I hope now that we can all open our eyes, see the beauty of the world that is in front of us, and shift our behaviors to be more in line with what will protect our own health and the health of the environment.

You absolutely can do it. I know you can. And you can start today. Right now.

APPENDIX
Recipes You Can Use

Recipe 1: Eat More Plants (and Significantly Less Meat)!

Prep: 15–20 minutes   Yield: A healthier life with less chronic disease

Ingredients:

- Pile each plate you create high with vegetables, whole grains, legumes, nuts, and seeds.
- Eat fresh, frozen, or canned fruit and vegetables.
- Gradually reduce your portion of meat, use it as a "flavoring" if you are not ready to give it up, or substitute meat alternatives (Impossible Meat, Beyond Meat, seitan).
- Substitute nondairy milks (coconut, soy, almond, pea) and yogurts for dairy products.

Directions:

- Don't be afraid of new flavors.
- Try out new (uncomplicated) recipes.
- Create a grocery list to help you shop.
- Give it a go!

Recipe 2: Buy Organic or Fair Trade

Prep: 15–20 minutes	Yield: Protect your health by eating produce grown without harsh petrochemical pesticides and fertilizers and support the environment

Ingredients:

- Meal planning chart.
- Grocery list.
- Search engine to learn more about various companies.

Directions:

- Look at your seven-day meal plan so that you know what you will need and when.
- Use more perishable items first or cook with them to make them last longer.
- Use the meal plan to create a shopping list so that you know how much to buy.
- Properly store foods in the pantry, refrigerator, or freezer so that they do not need to be discarded.

Recipe 3: Eat More Omegas (but Ditch the Fish)!

Prep: 5–10 minutes	Yield: Healthy essential fatty acids without a fishy taste or the risk of heavy metals or plastic by-products

Ingredients:

- Plant-based omega-3 fatty acids, including:
  - o Chia seeds
  - o Flaxseeds
  - o Walnuts
  - o Brussel sprouts
  - o Algal oil
  - o Hempseed

Directions:

- Sprinkle seeds or walnuts on top of cereals, salads, nondairy yogurts, or toast and jam.
- Soak chia seeds in nondairy milk and add flavorings to create a "pudding."
- Broil or roast brussels sprouts in good olive oil and add grapes and herbs for a delicious, juicy, and deep flavor.

Recipe 4: Decrease the Amount of Food You Waste

Prep: 15–20 minutes	Yield: More money in your bank account and less food thrown out

Ingredients:

- Meal planning chart.
- Grocery list.

Directions:

- Create a seven-day meal plan so that you know what you will need and when.
- With this meal plan, create meals that will use perishables early in the week and longer-lasting foods later in the week.
- Use this meal plan to create a shopping list so that you know how much to buy.
- Properly store foods in the pantry, refrigerator, or freezer so that they do not need to be discarded.

Recipe 5: Buy in Bulk to Reduce Packaging

Prep: 15–20 minutes	Yield: Reusable fabric grocery bags, money saved, and less plastic waste

Ingredients:

- Meal planning chart.
- Grocery list.
- Reusable produce or grocery bags.

Directions:

- Identify what food ingredients can be purchased in bulk and stored properly in the pantry, refrigerator, or freezer for long-term use. Good examples include:
  o Beans
  o Peas or lentils
  o Oats
  o Dried fruits
  o Pasta, rice, and other uncooked grains
- Create meal plans to help you identify what you will need and when.
- Seek out recipes that utilize some of these bulk ingredients.

Recipe 6: Can, Jar, or Freeze

Prep: 2–3 hours	Yield: Several months to years of inexpensive, healthy, preserved foods that can be used in case of emergency or for improved longevity of foods

Ingredients:

- Airtight, sealable, freezable containers.
- Glass mason jars appropriate for canning and jarring.
- A good website link with instructions for canning and jarring various foods.

Directions:

- Identify some of your favorite foods that are good candidates for freezing, canning, or jarring.
- Good foods that can be frozen long term include:
  o Nuts, flours, and breads
  o Herbs
  o Berries, peeled bananas, and other fruits (can be used in pancakes or smoothies later)
  o Blanched vegetables
  o Soups and other prepared mixed dishes (e.g. vegetable lasagna)
- Good foods that can be canned/jarred long term include:
  o Fruits and fruit juices
  o Jams and jellies, chutneys, sauces, and pie fillings
  o Salsas, tomatoes, and certain other vegetables
  o Pickles, relishes, vinegars, or other condiments
  o Chili and soups

Recipe 7: Compost

Prep: Varies depending on the method	Yield: Recycled soil nutrients, healthier gardens, and less food waste diverted to landfill

Ingredients:

- Home composting location.
- City-provided composting bin.

Directions:

- Place all food discards into a composting bin (either at home or city-provided) rather than in the rubbish/landfill pile. Follow compositing instructions.

Recipe 8: How to: Home and Community Gardens

Prep: Initial setup and planting	Yield: Homegrown and delicious fruits and vegetables for harvesting, eating, storing, replanting, and minimizing food mileage

Ingredients:

- Raised beds.
- Soil.
- Seeds.
- Sweat.
- Shovels, rakes, water, tender loving care (TLC).

Directions:

- Review what types of seeds grow best during each season and in each latitude and climate.
- Purchase seeds for that season and read the planting instructions on the packet.
- Give some water, sunshine, and time.
- Harvest.
- Replant and repeat.

Recipe 9: When You Cannot Grow It Yourself, Shop at Local Farmers Markets and Use Community-Supported Agriculture!

Prep: 30 minutes to 1 hour	Yield: Delicious farm-fresh fruits and vegetables from your local farm and lower food mileage

Ingredients:

- List of local farmers markets.
- Reusable bags.

Directions:

- Come with an open mind and several reusable produce or grocery bags to scoop up anything that looks delicious.
- Come with questions – farmers typically enjoy answering questions about their farms and their produce.
- Seek out delicious recipes for some of the new produce you may be trying out for the first time.

Recipe 10: Reduce, Reuse, and Refuse the Plastic: There Are Better Alternatives!

Prep: 2–3 minutes   Yield: Less plastic waste introduced into the environment

Ingredients:

- Reusable grocery bags.
- Reusable coffee mug.
- Stainless steel or glass water bottle.
- Conviction.

Directions:

- When asked about whether you want a plastic bag to carry items, kindly decline or pull out your own bag.
- Bring your own reusable straw or kindly decline a plastic straw to drink with.
- Use your own reusable coffee mug or water bottle.

Recipe 11: Seek Out Compostable and Eco-Friendly Products

Prep: 15–20 minutes   Yield: Reduced petrochemical plastic in landfills and in the oceans and safer and healthier products and containers

Ingredients:

- List of biodegradable, compostable, and eco-friendly products and manufacturers.
- Search engine to learn more about various companies.

Directions:

- Look at the labels of the products you buy and seek out specific compostable or biodegradable items, which are often printed with a specific label or number.
  o Example: ASTM D6400 (USA) or D8686.
  o CEN and ISO are European equivalents.
- If making an online purchase, search for biodegradable and compostable products (this is often easier than going through several shelves of products in a store).

Recipe 12: Engage in Sustainable or Ecological Tourism

Prep: Hours, Days, or Weeks	Yield: Most often a better experience that gives back to the community and helps the animals and habitat being visited

Ingredients:

- Possibly a travel agent.
- List of eco-friendly and ecotourism websites.
- Search engine to learn more about various companies.

Directions:

- Seek out experiences you want to engage in through ecotourism or sustainable tourism companies.
- Plan your trip and inquire as to what the experience will be like.
- Take many pictures and hold many memories.
- Have fun and give back to the community you are engaging with.

Recipe 13: Buy Fewer Things and Less "Stuff"

Prep: Variable	Yield: Improved sense of well-being, especially when engaging in experiences rather than purchasing "things"

Ingredients:

- Plan experiences rather than buying things.
- You and your loved ones.

Directions:

- Think about what experiences you would like to partake in.
- Think about what you really need or want and create a list.
- Decide what you want to do.
- Make a plan and do it.
- Avoid buying "things" for a fleeting "better" feeling that probably will not last.

Recipe 14: Engage with Your State and Local Officials

Prep: Minutes to hours	Yield: Better local programs that will benefit you, the people you care about, and the environment

Ingredients:

- Ideas.
- Email addresses of local leaders.
- City and county websites with "contact us" pages.

Directions:

- Bring concerns you have about your community to your local leaders through their "contact us" pages, phone numbers, offices, or email addresses.
- Attend local district meetings.
- Join or attend groups that work on issues that are important to you.
- Advocate for or work with your local city council on environmental issues.
- Know that you can make a difference.

Recipe 15: Vote for and Elect Public Officials Who Believe in and Support Climate Change and Eco-Friendly Policies

Prep: Variable	Yield: Improved environmental policies and laws that protect the world in which we live for ourselves, our children, and our grandchildren

Ingredients:

- Understanding of your candidate's background and policy positions.
- Civic duty.

Directions:

- Vote.
- Vote.
- Vote.
- If there is no perfect "environmental" candidate, still go and vote for the "least worst" candidate! (It may at least be a step in a better direction.)

Recipe 16: Walk, Bike, or Use Public or Shared-Ride Transit When Possible and Advocate for More Walkable Cities

Prep: Depends on distance	Yield: Improved personal health and well-being through increased exercise and better air quality

Ingredients:

- Body.
- Shoes.
- Push-pedal bicycle (optional).

Directions:

- Plot your distance and decide on the safest method of transport.
- Walk, bike, or use public/shared-ride transit (may require a face covering).
- Breathe cleaner air and feel healthier!

Recipe 17: Use Less Electricity (and Energy) and Use Sustainable Sources Where Possible

Prep: Several hours/days	Yield: Lower electricity bills and lower carbon footprint

Ingredients:

- LEDs or CFLs.
- Window insulation.
- A packed freezer.
- Solar or photovoltaic panels if possible.
- Change your thermostat settings to be less energy consuming.

Directions:

- Check government websites for discounts on photovoltaic panels/installation (rebates or subsidies).
- Install DIY window coverings and insulation or hire someone to help.
- Pack your freezer full to use less energy.
- Lower your thermostat in the winter and raise it in the summer.

Recipe 18: Educate Others! See, Do, Teach, and Lead by Example

Prep: None needed	Yield: Improved self-efficacy and self-esteem knowing that you are doing all you can to improve your health, the health of others, and the health of the environment

Ingredients:

- Your voice.
- Your knowledge.
- Your actions.
- Leadership.

Directions:

- Talk to and educate others if they ask you about sustainability and the environment.
- Teach them what you know and/or point them to this book for additional information.
- Lead by example and take charge of and engage in sustainable activities.
- Do the things that keep you healthy and help the environment.

Recipe 19: Plant Trees and Other Plants to Protect the Environment

Prep: Several hours	Yield: Cleaner air, cooler temperatures, more oxygen, and better psychological well-being

Ingredients:

- Trees.
- Plants.
- Soil.
- Water.

Directions:

- Engage with city planners to plant more trees within your city, county, district, etc.
- At your own home, plant more trees and other greenery, especially shade trees.
- Spend time in nature and outdoors with trees and other greenery.

Recipe 20: Seek Out Cotton or Other Natural Fibers, Not Synthetics

| Prep: Minutes to hours | Yield: Improved skin health as natural fibers allow skin to breathe, reduce itchiness and irritation, and are usually better for the environment |

Ingredients:

- Cotton fibers.
- Hemp fibers.
- Silk fibers.
- Linen fibers.

Directions:

- Carefully look at clothing and other fabric labels to see what they are made of.
- Seek out natural materials/ingredients as listed above.
- Limit your use of polyester, rayon, nylon, acrylic, or spandex.
- Learn more about the ethics of the company you are buying from.
- Wash products carefully by hand or in a front-loading washing machine.
- Enjoy.

Recipe 21: Read These Books and Watch These Films

| Prep: Hours | Yield: Increased knowledge, entertainment, and feeling more engaged with these topics |

Ingredients:

- Books listed in Chapter 8.
- Films/documentaries listed in Chapter 8.

Directions:

- Read each book and then seek out some more you find interesting.
- Watch each film and then seek out some more you find interesting.
- Take notes (optional).
- Discuss with others (optional).

NOTES

1 WFP, 2020a.
2 CDC, 2005.
3 CDC, 2019b.
4 WHO, 2019a.
5 Ji, 2020.
6 Quammen, 2013.
7 IEA, 2020.
8 Edmond, 2020.
9 Reuters, 2020.
10 Troeng, Barbier, & Rodriguez, 2020.
11 Londono, Andreoni, & Casado, 2020.
12 Easterling et al., 2007.
13 IPCC, 2007.
14 Denman et al., 2007.
15 Smith & Meade, 2019.
16 Boko et al., 2007.
17 WHO, 2018.
18 Cardona et al., 2012.
19 WFP, 2019.
20 Matthews, 2020.
21 Conservation International, 2020.
22 WHO, 2020c.
23 USDA-ERS, 2016.
24 USDA, 2003.
25 Zhong, Horn, & Cornelis, 2019.
26 Campbell & Campbell, 2016.
27 WHO/FAO, 2002.
28 WHO, 2016b.
29 Gerber et al., 2013.
30 White & Hall, 2017.
31 FAO, 2006.
32 EPA, 2016b.
33 Goodland & Anhang, 2009.

34 Springmann et al., 2016.
35 FAOSTAT, 2017.
36 FAO, 2011.
37 FAO, 2013a.
38 WHO, 2018.
39 Whiting et al., 2019.
40 Harvard School of Public Health, 2016.
41 Rice et al., 2000.
42 Dunn-Emke, Weidner, & Ornish, 2001.
43 Slavikova, 2019.
44 Siewerski, 2020.
45 Liu, Kuchma, & Krutovsky, 2018.
46 Harun & Ogneva-Himmelberger, 2013.
47 Gurian-Sherman, 2008.
48 Samuel, 2020.
49 Anthis, 2019.
50 Martin, Thottathil, & Newman, 2015.
51 Boeckel et al., 2015.
52 CDC, 2017.
53 Patton, 2020.
54 Hollenbeck, 2016.
55 WHO, 2020a.
56 WHO, 2019b.
57 Banerjee et al., 2019.
58 Bintsis, 2017.
59 CDC, 2019a.
60 Worldometer, 2020.
61 UN-DESA, 2019.
62 Diamond, 2019.
63 UN, 2020.
64 Ritchie, 2019c.
65 World Population Review, 2020.
66 UN-DESA, 2018.
67 WHO-Euro, 2020.
68 FAO, 2000.
69 Hunnes, 2013.
70 Francis & Francis, 2006.
71 Earth Overshoot Day, 2020a.
72 FAO, 2018a.
73 Butler, 2019.
74 IFPRI, 2008.
75 Latawiec et al., 2017.
76 NASA, 2007.
77 Droogers, Seckler, & Makin, 2001.
78 Ritchie & Roser, 2017.
79 Shepon et al., 2016.
80 Berners-Lee et al., 2018.
81 FAO & IWMI, 2017.
82 Pimentel & Pimentel, 2003.
83 Koneswaran & Nierenberg, 2008.
84 Foer, 2020.

85 WFP, 2020b.
86 Estes, 2020.
87 Perch-Neilsen, Battig, & Imboden, 2008.
88 Yaffe-Bellany & Corkery, 2020.
89 FAO, 2020d.
90 World Economic Forum, 2020.
91 Koneswaran & Nierenberg, 2008.
92 Harvard HSPH, 2020.
93 Roser-Renouf, 2016.
94 Earth Overshoot Day, 2020b.
95 UN-DESA, 2015.
96 Foot Print Network, 2017.
97 Johns Hopkins University Center for a Livable Future, 2016.
98 USDA & DHHS, 2015b.
99 FAO, 2010.
100 FNB et al., 2015.
101 Herforth et al., 2019.
102 AHA, 2015.
103 European Commission, 2020b.
104 FAO, 2020c.
105 Ton et al., 2016.
106 Mishra et al., 2013.
107 Siscovick et al., 2017.
108 NOAA, 2014.
109 Earle, 2010.
110 ISC, 2016.
111 FAO, 2018c.
112 WWF, 2020b.
113 FAO, 2011.
114 Valencia, 2020.
115 Clover, 2006.
116 Myers, Hutchings, & Barrowman, 1997.
117 Kelleher, 2005.
118 UNEP, 2016a.
119 Herr & Galland, 2009.
120 Milazzo et al., 2016.
121 Pinones & Fedorov, 2016.
122 NOAA, 2020a.
123 Rochman et al., 2015.
124 Eriksen et al., 2014.
125 Tilman & Clark, 2015.
126 Harari, 2015.
127 FAO, 2017b.
128 Okin, 2017.
129 Mottet et al., 2017.
130 Pistollato & Battino, 2014.
131 Satija et al., 2016.
132 Hawkins & Sabate, 2013.
133 Nestle, 2016.
134 Anhang, 2018.
135 Reinicke, 2019.

136 FAO & AQUASTAT, 2020.
137 WWAP, 2016.
138 DESA, 2011.
139 Savage, 2020.
140 DGAC, 2015.
141 WWF, 2018.
142 Ceballos, Ehrlich, & Dirzo, 2017.
143 McRae, Deinet, & Freeman, 2016.
144 Normile, 2016.
145 UNEP, 2018.
146 Branch et al., 2013.
147 McManus et al., 2019.
148 Diaz & Rosenberg, 2008.
149 EAS, 2009.
150 Sprague, Dick, & Trocher, 2016.
151 Roach, 2008.
152 FAO, 2005b.
153 World Bank, 2013.
154 Bale, 2016.
155 Cole et al., 2009.
156 Behringer et al., 2020.
157 Folke & Kautsky, 1992.
158 APCSS, 1998.
159 Cribb, 2019.
160 Gleick & Cooley, 2009.
161 Earth Day, 2018.
162 Statista, 2017.
163 House of Commons Environmental Audit Committee, 2017.
164 UNRIC, 2013.
165 EPA, 2015.
166 Feinkel, 2011.
167 SPI, 2016b.
168 Plastics Europe, 2013.
169 Ritchie & Roser, 2018.
170 United States Census Bureau, 2016.
171 Allsopp et al., 2006.
172 Gourmelon, 2015.
173 European Commission, 2011.
174 Jambeck et al., 2015.
175 Earth Institute, 2012.
176 EPA, 2016a.
177 Bendix, 2019.
178 Earth Policy Institute, 2014.
179 Environment California, 2016.
180 GPI, 2016.
181 The Aluminum Association, 2016.
182 Sun, 2014.
183 US Fish and Wildlife Service, 2013.
184 Li, Tse, & Fok, 2016.
185 Cho, 2012.
186 UNEP, 2014.

187 Webb et al., 2013.
188 National Geographic, 2009.
189 UNEP, 2015.
190 Macfayden, Huntington, & Cappell, 2009.
191 NOAA, 2016.
192 FAO, 2018b.
193 Fukuoka et al., 2016.
194 Lewis, 2019.
195 NOAA, 2008.
196 UNEP, 2016c.
197 Earth Institute, 2011.
198 Rochman et al., 2015.
199 UNEP, 2013.
200 Lin, 2016.
201 EPA, 2017a.
202 Teuten et al., 2009.
203 Alexander, 1999.
204 Wong et al., 2013.
205 Krahn et al., 2009.
206 Lundin et al., 2016.
207 Lundin et al., 2015.
208 Lloret et al., 2016.
209 Guo et al., 2019.
210 ATSDR, 2012.
211 Calafat, 2008.
212 Galloway et al., 2017.
213 Kelland, 2010.
214 Ozen & Darcan, 2011.
215 Ellahi & Rashid, 2018.
216 Petre, 2018.
217 McIntyre & Frohlich, 2015.
218 SPI, 2015.
219 Statista, 2016.
220 Mersha et al., 2015.
221 WHO, 2014.
222 Sax, 2010.
223 CPSC, 2015.
224 NIEHS–NIH, 2016.
225 EWG, 2013.
226 CITES, 2013.
227 CITES, 2020.
228 FWS, 2015.
229 Moyle, 2009.
230 Oceana, 2020.
231 Shiffman, 2020.
232 Roff et al., 2018.
233 *Geographical Magazine*, 2006.
234 Biggs et al., 2013.
235 Kolbert, 2014.
236 Africa Geographic, 2014.
237 Chase et al., 2016.

238 Angier, 2018.
239 Steyn, 2016.
240 Maron, 2018b.
241 Compaore et al., 2020.
242 Shaffer et al., 2019.
243 King et al., 2017.
244 O'Connell-Rodwell, 2010.
245 McComb et al., 2011.
246 Colbeck, 2010.
247 Bradshaw, 2009.
248 Safina, 2015.
249 Hsu, 2017.
250 World Atlas, 2019.
251 Bradford, 2018.
252 Nuwer, 2018.
253 *Nature*, 1938.
254 Morell, 2015.
255 Daley, 2018.
256 GCF, 2018.
257 Neme, 2010.
258 Endangered Species Journalist, 2020.
259 Maron, 2018a.
260 WWF, 2020a.
261 Frost, 2020.
262 NHES, 2020.
263 Yan, 2020.
264 EIA, 2020.
265 Dasgupta, 2018.
266 Mexicanist, 2020.
267 Yifan, 2018.
268 Marino, 2020.
269 Cowperthwaite, 2013.
270 Shepherdson, Carlstead, & Wielebnowski, 2004.
271 Jett & Ventre, 2015.
272 Lahdenpera et al., 2018.
273 Russo, 2015.
274 Platt, 2015.
275 Jaramillo-Legorreta et al., 2019.
276 Rojas-Bracho et al., 2019.
277 Ladkani, 2019.
278 IPBES, 2019.
279 Anderson, 2018.
280 Johnson et al., 2017.
281 Wilcox, 2018.
282 De Vos et al., 2015.
283 Davis, Faurby, & Svenning, 2018.
284 Yong, 2018.
285 Ritchie, 2019b.
286 Bar-On, Phillips, & Milo, 2018.
287 Zeller, Starik, & Gottert, 2017.
288 GRAIN & IATP, 2018.

289 Lolu et al., 2020.
290 Hermes et al., 2018.
291 Barnosky & Hadly, 2015.
292 Mendoza et al., 2020.
293 Forgey, 2020.
294 Lu et al., 2016.
295 Yang et al., 2020.
296 Flack, 2016.
297 Marino et al., 2020.
298 Dalton, 2019.
299 WDC, 2020.
300 Pollard, 2019.
301 Orca Network, 2020.
302 Safiq, 2018.
303 Dolphin Project, 2020b.
304 Curtin & Wilkes, 2007.
305 Psihoyos, 2009.
306 World Animal Protection, 2020.
307 Thornton, 2000.
308 Reiss, 2011.
309 JCU, 2020.
310 Thailand Elephants, 2020.
311 Daly, 2019.
312 Iyer, 2016.
313 Sheldrick Wildlife Trust, 2020.
314 issuu, 2015.
315 DSWT, 2016.
316 DSWT, 2020.
317 Khatchadourian, 2007.
318 Heller, 2006.
319 Beer, 2018.
320 Sea Shepherd, 2020.
321 Watson, 2015.
322 Sullivan, 2020.
323 CDP, 2020.
324 Davidson, 2020.
325 Locker, 2019.
326 Latif, 2020.
327 Serna, 2020.
328 Borys, 2020.
329 Vally, 2020.
330 ASM, 2020.
331 Aymerich-Franch, 2020.
332 Peters et al., 2019.
333 UNEP, 2019.
334 UNFCCC, 2015.
335 Jackson, 2020.
336 Colarossi, 2020.
337 Block, 2020.
338 Reid, 2020.
339 WHO, 2020b.

340 Macias, 2020.
341 Mastroianni, 2020b.
342 Valdivia & Margolies, 2020.
343 4ocean, 2020.
344 IAPF, 2020.
345 Tikki Hywood Foundation, 2020.
346 PAWS, 2020.
347 Dolphin Project, 2020a.
348 O'Barry, 2020.
349 International Animal Rescue, 2020.
350 Lewis Pugh Foundation, 2020.
351 Knapp, Peace, & Bechtel, 2017.
352 Handmer et al., 2012.
353 Maxwell & Caldwell, 2008.
354 Lavell et al., 2012.
355 Deressa, Ringler, & Hassan, 2010.
356 Psihoyos, 2015.
357 Wu, 2016.
358 O'Malley, Lee-Brooks, & Medd, 2013.
359 Platt, 2014.
360 Buckley, Morrison, & Castley, 2016.
361 Gallagher & Hammerschlag, 2011.
362 Aragona & Orr, 2011.
363 UN, 2015a.
364 Wynes & Nicholas, 2017.
365 Ontl & Schulte, 2012.
366 Waite & Rudee, 2020.
367 Tickell, 2017.
368 UNSCN, 2018.
369 WHO, 2016a.
370 Vuuren, 2017.
371 CIA, 2016.
372 Purdy, 2013.
373 Yang et al., 2008.
374 Hawkins, 2019.
375 Psihoyos, 2018.
376 UN FAO, 2006.
377 WHO, 2020d.
378 Tuso et al., 2013.
379 Esselstyn, 2007.
380 Toumpanakis, Turbull, & Alba-Barba, 2018.
381 Branch, 2020.
382 Phalan et al., 2016.
383 FAO, 2005a.
384 Bowling, 2015.
385 Fothergill, Hughes, & Scholey, 2020.
386 Marlow et al., 2014.
387 FAO, 2016a.
388 WFP, 2016.
389 McArthur & Rasmussen, 2016.
390 Tirado et al., 2015.

391 GFN, 2016.
392 USDA & DHHS, 2015a.
393 Makowski, 2019.
394 Simon, Davies, & Ancrenaz, 2019.
395 CORIS, 2020.
396 Kim, Bang, & Lee, 2019.
397 Science Daily, 2015.
398 Guibourg & Briggs, 2019.
399 Harvard Medical School, 2008.
400 Beckett & Oltjen, 1993.
401 Ethical Consumer, 2020.
402 Campion et al., 2020.
403 Mederios et al., 2020.
404 Hughes, 2017.
405 Martinez et al., 2010.
406 Fair Trade Certified, 2020.
407 Bartram & Perkins, 2003.
408 Seufert, Ramankutty, & Foley, 2012.
409 Bruggen et al., 2018.
410 Associated Press, 2019.
411 Kim et al., 2019.
412 Meemken et al., 2019.
413 Grey & Bolland, 2014.
414 AHA, 2016.
415 WebMD, 2018.
416 Reid & Budge, 2014.
417 Ryan et al., 2009.
418 Sprague, Betancor, & Tocher, 2017.
419 Finco et al., 2017.
420 Lenihan-Geels, Bishop, & Ferguson, 2013.
421 Peltomaa, Johnson, & Taipale, 2018.
422 Brenna, 2002.
423 Burdge, 2006.
424 NIH–ODS, 2019.
425 LPI–OSU, 2019.
426 Troell et al., 2009.
427 Voort et al., 2017.
428 Paris, 2019.
429 SFP, 2018.
430 Adhikari, Barrington, & Martinez, 2006.
431 Jribi et al., 2020.
432 Feeding America, 2020.
433 Food Foundation, 2016.
434 Stuart, 2012.
435 Khokhar, 2013.
436 UN, 2019.
437 Gunders, 2012.
438 FAO, 2020b.
439 Reardon, Bellemare, & Zilberman, 2020.
440 FAO, 2016b.
441 Clarke, Schweitzer, & Roto, 2015.

442 WRAP, 2020.
443 EEA, 2016.
444 UN Environment, 2020.
445 EPA, 2016e.
446 European Commission, 2020a.
447 Eveleth, 2014.
448 IFT, 2016.
449 NPR, 2016.
450 NPR, 2014.
451 Parker, 2018.
452 Vann, 2020.
453 Davis, 2014.
454 Sass, 2014.
455 Muncke et al., 2014.
456 Trasande, Shaffer, & Sathyanarayana, 2018.
457 Neltner et al., 2011.
458 Mastroianni, 2020a.
459 Klein, 2020.
460 Teixeira, 2015.
461 USDA, 2015.
462 Sultana, Anwar, & Iqbal, 2008.
463 FAO, 2003.
464 Bryant, 2015.
465 Mehta et al., 2014.
466 Palansooriya et al., 2020.
467 Sage Publications, 2008.
468 CARP & CEPA, 2011.
469 Cal Recycle, 2020.
470 Brown, 2014.
471 Gerlock, 2014.
472 Department of Primary Industries and Regional Development, 2018.
473 UNL, 2020.
474 USCC, 2008.
475 Logan, 2019.
476 EPA, 2019.
477 Simon, 2020.
478 University of Wyoming, 2020.
479 Managa et al., 2018.
480 Lussier, 2018.
481 Yale Environment Review, 2017.
482 Harris, 2009.
483 Feldman, 2015.
484 Marin Master Gardeners, 2016.
485 Ferrin, Norman, & Sempik, 2001.
486 Askew, 2018.
487 IFAD, 2019.
488 Apetkar & Myers, 2020.
489 Heim, Stang, & Ireland, 2009.
490 Aguilar-Stoen, Moe, & Carmago-Ricalde, 2009.
491 Tarkan, 2015.
492 USDA NRCS, 2009.

493 USDA, 2020.
494 Veldstra, Alexander, & Marshall, 2014.
495 European Commission, 2019.
496 Holland, 2020.
497 Cone & Myhre, 2000.
498 Adam, 2006.
499 Cox et al., 2008.
500 FoodPrint, 2020.
501 Vasquez et al., 2017.
502 Christensen, Galt, & Kendall, 2018.
503 Galt et al., 2017.
504 Danovich, 2020.
505 Yaroslavsky et al., 2007.
506 European Commission, 2017.
507 ScAAN, 2019.
508 National Conference of State Legislatures, 2020.
509 Ivanova, 2019.
510 Tenenbaum, 2020.
511 Henderson & Green, 2020.
512 Eriksen et al., 2014.
513 World Economic Forum, Ellen MacArthur Foundation, & McKinsey & Company, 2016.
514 Harrington, 2017.
515 EarthDay.org, 2018.
516 GreenMatch, 2019.
517 Moss, 2015.
518 Food and Water Watch, 2020.
519 Gibbens, 2019.
520 Westerhoff et al., 2008.
521 LA Times Editorial Board, 2018.
522 Abilash & Sivapragesh, 2013.
523 Iwata, 2015.
524 BPI, 2020.
525 National Geographic, 2009.
526 ASTM International, 2020.
527 Gibbens, 2018.
528 Khare & Deshmukh, 2006.
529 Neilsen et al., 2017.
530 Essential Chemical Industry, 2013.
531 Flint, 2013.
532 Kinhal, 2020.
533 West, 2019.
534 World Centric, 2018.
535 Washam, 2010.
536 Royte, 2006.
537 Mostafa et al., 2018.
538 Ach, 2006.
539 Wagner, 2014.
540 US-EIA, 2020.
541 MChemical, 2020.
542 Abdel-Motaal et al., 2014.

543 Rudnik, 2013.
544 Lackner, 2016.
545 Ferreira et al., 2017.
546 Baker et al., 2012.
547 Oever et al., 2017.
548 Sehgal & Singh, 2010.
549 TIES, 2019.
550 The World Counts, 2020.
551 Anderson & Altmann, 2019.
552 Brymer & Lacaze, 2013.
553 Chiu, Lee, & Chen, 2014.
554 Curtin, 2009.
555 Rashid, 2013.
556 Ahuvia, 2008.
557 Maslow, 1943.
558 Pchelin & Howell, 2014.
559 Kumar, Killingsworth, & Gilovich, 2020.
560 Hamblin, 2014.
561 Pozin, 2016.
562 Kumar, Killingsworth, & Giilovich, 2014.
563 Helm et al., 2019.
564 Portney, 2005.
565 EDF, 2011.
566 Romer & Tamminen, 2014.
567 Behavioral Economics, 2020.
568 Sawe, 2019.
569 Biermann & Boas, 2010.
570 Naser et al., 2019.
571 Newell, 2005.
572 Samet & Burke, 2020.
573 Montgomery, 2017.
574 McAuley & Spolar, 2020.
575 Leahy, 2019.
576 Johnson, 2020.
577 Pearce, 2019.
578 Lenton et al., 2020.
579 Ge & Friedrich, 2020.
580 EPA, 2020b.
581 Wang & Ge, 2019.
582 USA Facts, 2019.
583 EPA, 2020a.
584 IPCC, 2014.
585 Quéré et al., 2020.
586 Woodyatt, 2020.
587 Andrew, 2020.
588 Borenstein, 2020.
589 IATA, 2020.
590 Quéré et al., 2020.
591 Vandy, 2020.
592 Perry, 2020.
593 McAuley & Spolar, 2020.

594 CDC, 2020.
595 House of Commons Library, 2019.
596 WHO-Euro, 2020.
597 Berman, Jonies, & Kaplan, 2008.
598 Payam et al., 2015.
599 Khan & Arsalan, 2016.
600 Jones, 1991.
601 Kemp, 2019.
602 Earl, 2020.
603 GWEC, 2016.
604 McCully, 1996.
605 USFWS, 2020.
606 Con, 2020.
607 IRENA, 2019.
608 Desta, 2014.
609 SWPPD, 2020.
610 Montoya et al., 2017.
611 Castro, 2013.
612 US Department of Energy, 2012.
613 EIA, 2013.
614 European Commission, 2020c.
615 Carrero et al., 2020.
616 Kahleova, Levin, & Barnard, 2018.
617 Madigan & Karhu, 2018.
618 Spaderna et al., 2013.
619 Brown, 2020.
620 Hale, 2018.
621 Walch, 2020.
622 Vidyasagar, 2018.
623 FAO, 1992.
624 Felton & Smith, 2017.
625 Turner, 2019.
626 McPherson & Rowntree, 1993.
627 Rudee, 2020.
628 Dwyer, Schroeder, & Gobster, 1991.
629 White et al., 2019.
630 Stancil, 2019.
631 Nowak, Hoehn, & Crane, 2007.
632 NOAA, 2020b.
633 National Geographic, 2020b.
634 Zimmer, 2019.
635 FAO, 2020a.
636 Royal Parks, 2020.
637 Sastry, 2018.
638 Wicker, 2017.
639 Changing Markets Foundation, 2019.
640 Green Choices, 2020.
641 Chen & Burns, 2006.
642 Advameg, 2020.
643 Fashion Materials, 2020.
644 Hartline et al., 2016.

645 Plastic Soup Foundation, 2018.
646 Napper & Thompson, 2016.
647 MERMAIDS EU Life+, 2016.
648 Resnick, 2019.
649 Browne et al., 2011.
650 Wieczoreket al., 2018.
651 Obbard et al., 2014.
652 Kosuth, Mason, & Wattenberg, 2018.
653 IUCN, 2017.
654 Business Ethics, 2010.
655 Hymann, 2019.
656 Islam et al., 2020.
657 Guillingsrud, 2017.
658 1 Million Women, 2016.
659 Global Commodities-Silk, 2020.
660 Allen, 2019.
661 GHG Online, 2019.
662 Griplas, 2019.
663 AZoM, 2012.
664 Gould, 2015.
665 BBC, 2020.
666 Thornton, 2019.

BIBLIOGRAPHY

1 Million Women (2016). *How to compost fabrics*. Retrieved from 1 Million Women: www.1millionwomen.com.au/blog/how-compost-fabrics/

1bag at a time (2020). *Plastic bags and petroleum*. Retrieved from 1bag at a time: https://1bagatatime.com/learn/plastic-bags-petroleum/

4ocean (2020). *About us*. Retrieved from 4ocean: https://4ocean.com/about/

Abdel-Motaal, F., El-Sayed, M., El-Zayat, S., & Ito, S. (2014). Biodegradation of poly (ε-caprolactone) (PCL) film and foam plastic by *Pseudozyma japonica* sp. nov., a novel cutinolytic ustilaginomycetous yeast species. *3 Biotech*, 4, 507–512.

Abilash, N. & Sivapragesh, M. (2013). Environmental benefits of eco-friendly natural fiber reinforced polymeric composite materials. *International Journal of Application or Innovation in Engineering & Management*, 2, 53–59.

Ach, A. (2006). Biodegradable plastics based on cellulose acetate. *Journal of Micromolecular Science*, 30, 733–740.

Adam, K. L. (2006). *Community Supported Agriculture*. Butte, MT: ATTRA, National Sustainable Agriculture Information Service.

Adhikari, B. K., Barrington, S., & Martinez, J. (2006). Predicted growth of world urban food waste and methane production. *Waste Management & Research*, 24, 421–433.

Advameg (2020). *Acrylic plastic*. Retrieved from Made How: www.madehow.com/Volume-2/Acrylic-Plastic.html

Africa Geographic (2014). *When the buying stops, the killing can too*. Retrieved from African Geographic: http://africageographic.com/blog/when-the-buying-stops-the-killing-can-too/

Aguilar-Stoen, M., Moe, S., & Carmago-Ricalde, S. (2009). Home gardens sustain crop diversity and improve farm resilience in Candelaria Loxicha, Oaxaca, Mexico. *Human Ecology*, 27, 55–77.

AHA (2015). *Fish 101*. Retrieved from American Heart Association: www.heart
.org/HEARTORG/HealthyLiving/HealthyEating/Fish-101_UCM_305986_
Article.jsp#.V6J7WKKPaxg

AHA (2016). *Omega-3 fatty acids from fish oil may help healing after heart
attack*. Retrieved from American Heart Association: https://news.heart.org/
omega-3-fatty-acids-from-fish-oil-may-help-healing-after-heart-attack/

Ahuvia, A. (2008). If money doesn't make us happy, why do we act as if it does?
Journal of Economic Psychology, **29**, 491–507.

Albert, S., Leon, J., Grinham, A., Church, J., Gibbes, B., & Woodroffe, C.
(2016). Interactions between sea-level rise and wave exposure on reef
island dynamics in the Solomon Islands. *Environmental Research Letters*,
11, 054011.

Alexander, D. (1999). Bioaccumulation, bioconcentration, biomagnification. In
Environmental Geology. Encyclopedia of Earth Science. Dordrecht: Springer,
pp. 43–44.

Allen, E. (2019). *The environmental impact of wool*. Retrieved from The
Ecologist: https://theecologist.org/2019/mar/12/environmental-impact-wool

Allsopp, M., Walters, A., Santillo, D., & Johnston, P. (2006). *Plastic Debris in the
World's Oceans*. Amsterdam: Greenpeace for UNEP.

Anderson, K. (2018). *What's normal: how scientists calculate background extinction rate*. Retrieved from Population Education: https://populationeducation
.org/what-is-background-extinction-rate-how-is-it-calculated/

Anderson, W. & Altmann, M. (2019). The effects of wildlife-based ecotourism. In
J. Koprowski & P. Krausman (eds.), *International Wildlife Management*.
Baltimore, MD: Johns Hopkins University Press, chapter 9.

Andrew, S. (2020). *Covid-19 lockdowns could drop carbon emissions to their
lowest level since World War II. But the change may be temporary*. Retrieved
from CNN: www.cnn.com/2020/05/19/world/carbon-emissions-coronavirus-
pandemic-scn-climate-trnd/index.html

Angier, N. (2018). *How teeth became tusks, and tusks became liabilities*.
Retrieved from *New York Times*: www.nytimes.com/2018/09/11/science/
tusks-teeth-elephants-genes.html

Anhang, J. (2018). Re-thinking meat: how climate change is disrupting the meat
industry. In D. Bogueva, D. Marinova, & T. Raphaely (eds.), *Handbook of
Research on Social Marketing and Its Influence on Animal Origin Food
Product Consumption*. Hershey, PA: IGI Global, pp. 311–326.

Anthis, K. (2019). *Global farmed and factory farmed animals estimates*.
Retrieved from Sentience Institute: www.sentienceinstitute.org/global-animal-
farming-estimates#ftnt1

APCSS (1998). *Food Security and Political Stability in the Asia-Pacific Region*.
Honolulu, HI: Asia-Pacific Center for Security Studies.

Apetkar, S. & Myers, J. (2020). The tale of two community gardens: green aesthetic versus food justice in the Big Apple. *Agriculture and Human Values*, 37, 779–792.

Aragona, F. & Orr, B. (2011). Agriculture intensification, monocultures, and economic failure: the case of onion production in the Tipajara Watershed on the eastern slope of the Bolivian Andes. *Journal of Sustainable Agriculture*, 35, 467–492.

Askew, K. (2018). *Urban farming could reinvent supply chains*. Retrieved from Food Navigator: www.foodnavigator.com/Article/2018/08/08/Urban-farming-could-reinvent-supply-chains#

ASM (2020). *The science of social distancing*. Retrieved from American Society for Microbiology: https://asm.org/Articles/2020/April/The-Science-of-Social-Distancing

Associated Press (2019). *Good news if you buy organic food – it's getting cheaper*. Retrieved from Marketwatch.com: www.marketwatch.com/story/heres-why-prices-of-organic-food-are-dropping-2019-01-24

ASTM International (2020). *ASTM D6400 Standard specification for labeling of plastics designed to be aerobically composted in municipal or industrial facilities*. Retrieved from ASTM International: www.astm.org/search/fullsite-search.html?query=Compostable%20plastic

ATSDR (2012). *Public health statement for styrene*. Retrieved from Agency for Toxic Substances & Disease Registry – CDC: www.atsdr.cdc.gov/phs/phs.asp?id=419&tid=74

Aymerich-Franch, L. (2020). *COVID-19 lockdown: impact on psychological well-being and relationship to habit and routine modifications*. Retrieved from PsyArXiv: https://psyarxiv.com/9vm7r/

AZoM (2012). *How do you recycle acrylic resin?* Retrieved from AZO materials: www.azom.com/article.aspx?ArticleID=7945

Bacon, C. M., Getz, C., Kraus, S., Montenegro, M. & Holland, K. (2012). The social dimensions of sustainability and change in diversified farming systems. *Ecology and Society*, 17: 41.

Baker, M., Walsh, S., Schwartz, Z., & Boyan, B. (2012). A review of polyvinyl alcohol and its uses in cartilage and orthopedic applications. *Journal of Biomedical Materials Research Part B: Applied Biomaterials*, 100, 1451–1457.

Bale, R. (2016). *One of the world's biggest fisheries is on the verge of collapse*. Retrieved from National Geographic: www.nationalgeographic.com/news/2016/08/wildlife-south-china-sea-overfishing-threatens-collapse/

Banerjee, A., Kulcsar, K., Misra, V., Frieman, M., & Mossman, K. (2019). Bats and coronaviruses. *Viruses*, 11, 41.

Barnosky, A. D. & Hadly, E. A. (2015). *Tipping Point for Planet Earth: How Close Are We to the Edge?* London: William Collins Books.

Bar-On, Y., Phillips, R., & Milo, R. (2018). The biomass distribution on Earth. *PNAS*, **115**, 6506–6511.

Bartram, H. & Perkins, A. (2003). The biodiversity benefits of organic farming. In OECD, *Organic Agriculture Sustainability, Markets and Policies*. Wallingford: OECD, pp. 77–94.

BBC (2020). *Australia's fires "killed or harmed three billion animals."* Retrieved from BBC: www.bbc.com/news/world-australia-53549936

Beckett, J. & Oltjen, J. (1993). Estimation of the water requirement for beef production in the United States. *Animal Science*, **71**, 818–826.

Beer, J. (2018). *Marine animals are suffocating on plastic in this new Sea Shepherd PSA*. Retrieved from Fast Company: www.fastcompany .com/40557685/marine-animals-are-suffocating-on-plastic-in-this-new-sea-she pard-psa

Behavioral Economics (2020). Default (option/setting). Retrieved from Behavior Economics: www.behavioraleconomics.com/resources/mini-encyclopedia-of-be/default-optionsetting/

Behringer, D., Wood, C., Krkosek, M., & Bushek, D. (2020). Disease in fisheries and aqauculture. In *Marine Disease Ecology*. Oxford: Oxford University Press, pp. 183–210.

Bendix, A. (2019). *Cities could one day power homes and fuel cars with plastic, a giant step toward solving the pollution crisis*. Retrieved from Business Insider: www.businessinsider.com/plastic-waste-converted-fuel-electricity-2019-7

Berman, M., Jonies, J., & Kaplan, S. (2008). The cognitive benefits of interacting with nature. *Psychological Science*, **19**, 1207–1212.

Berners-Lee, M., Kennelly, C., Watson, R., & Hewitt, C. (2018). Current global food production is sufficient to meet human nutritional needs in 2050 provided there is radical societal adaptation. *Elementa: Science of the Anthropocene*, **6**, 52.

Bernstein, A., Ding, E., Willett, W., & Rimm, E. (2012). A meta-analysis shows that docosahexaenoic acid from algal oil reduces serum triglycerides and increases HDL-cholesterol and LDL-cholesterol in persons without coronary heart disease. *Journal of Nutrition*, **142**, 99–104.

Biermann, F. & Boas, I. (2010). Preparing for a warmer world: towards a global governance system to protect climate refugees. *Global Environmental Politics*, **10**, 60–88.

Biggs, D., Courchamp, F., Martin, R., & Possingham, H. P. (2013). Legal trade of Africa's rhino horns. *Science*, **339**, 1038–1039.

Bintsis, T. (2017). Foodborne pathogens. *AIMS Microbiology*, **3**, 529–563.

Block, I. (2020). *London, New York, Paris, and Milan give streets to cyclists and pedestrians*. Retrieved from Dezeen: www.dezeen.com/2020/05/07/london-new-york-paris-milan-cyclists-pedestrians/

Boeckel, T. V., Brower, C., Gilbert, M., Grenfell, B., Leven, S., Robinson, T., . . . Laxminarayan, R. (2015). Global trends in antimicrobial use in food animals. *PNAS*, 112, 5649–5654.

Boko, M., Niang, I., Nyong, A., Voge, C., Githeko, A., Medany, M., . . . Yanda, P. (2007). Climate change 2007: impacts, adaptation and vulnerability. Contribution of Working Group II to the Fourth Assessment Report of the Intergovernmental Panel on Climate Change. M. Parry, O. Canziani, J. Palutikof, P. V. Linden, & C. Hanson (eds.). Cambridge: Cambridge University Press, pp. 433–467.

Borenstein, S. (2020). *Study: world carbon pollution falls 17% during pandemic peak*. Retrieved from AP News: https://apnews.com/7c268d15e503eb9d46 doc35cd9ac3543?fbclid=IwARonuWiqCMAwuZi78roIqE8alWEyOhHkd FGF6sauXz254ffTpHaPsZI7LPQ

Borys, S. (2020). *Some of Australia's top firefighters are heading to California to face wildfires – and COVID-19*. Retrieved from ABC News Australia: www .abc.net.au/news/2020-08-27/california-asks-for-australian-help-to-battle-hun dreds-of-fires/12599796

Bowling, A. G. (2015). *A leading cause of everything: one industry that is destroying our planet and our ability to thrive on it*. Retrieved from *Stanford Environmental Law Journal (SELJ)*: https://journals.law.stanford.edu/stan ford-environmental-law-journal-elj/blog/leading-cause-everything-one-indus try-destroying-our-planet-and-our-ability-thrive-it

BPI (2020). *Confused by the Terms Biodegradable & Biobased?* New York: Biodegradable Products Institute.

Bradford, A. (2018). *Facts about rhinos*. Retrieved from LiveScience: www .livescience.com/27439-rhinos.html

Bradshaw, G. (2009). *Elephants on the Edge: What Animals Teach Us about Humanity*. Ann Arbor, MI: Sheridan Books.

Branch (2020). *Coping with the meat shortage*. Retrieved from Consumer Reports: www.consumerreports.org/healthy-eating/coping-with-the-meat-shortage/

Branch, T. A., DeJoseph, B., Ray, L., & Wagner, C. (2013). Impacts of ocean acidification on marine seafood. *Trends in Ecology & Evolution*, 28, 178–186.

Brenna, J. (2002). Efficiency of conversion of alpha-linolenic acid to long chain n-3 fatty acids in man. *Current Opinion in Clinical Nutrition and Metabolic Care*, 5, 127–132.

Brown, K. (2020). *The hidden toll of lockdown on rainforests*. Retrieved from BBC Future Planet: www.bbc.com/future/article/20200518-why-lockdown-is-harming-the-amazon-rainforest

Brown, L. (2014). *Food waste diversion is key to a sustainable community*. Retrieved from EPA Blog: https://blog.epa.gov/2014/12/10/food-waste-diver sion-is-key-to-a-sustainable-community/

Browne, M., Crump, P., Niven, S., Teuten, E., Tonkin, A., Galloway, T., & Thompson, R. (2011). Accumulation of microplastic on shorelines worldwide: sources and sinks. *Environmental Science & Technology*, **45**, 9175–9179.

Bruggen, A. V., He, M., Shin, K., Mai, V., Jeong, K., Finckh, M., & Morris, J. M., Jr. (2018). Environmental and health effects of the herbicide glyposate. *Science of the Total Environment*, **616–617**, 255–268.

Bryant, L. (2015). *Organic matter can improve your soil's water holding capacity*. Retrieved from Natural Resources Defense Council: www.nrdc .org/experts/lara-bryant/organic-matter-can-improve-your-soils-water-holding-capacity

Brymer, E. & Lacaze, A.-M. (2013). The benefits of ecotourism for visitor wellness. In R. Ballantyne, & J. Packer (eds.), *International Handbook on Ecotourism*. Cheltenham: Edward Elgar Publishing, pp. 217–232.

Buckley, R., Morrison, C., & Castley, J. (2016). Net effects of ecotourism on threatened species survival. *PLoS One*, **11**, e0147988.

Burdge, G. (2006). Metabolism of alpha-linolenic acid in humans. *Prostaglandins, Lekuotrienes, and Essential Fatty Acids*, **75**, 161–168.

Business Ethics (2010). *What's so bad about cotton?* Retrieved from Business Ethics: https://business-ethics.com/2010/08/07/1438-the-bad-side-of-cotton/

Butler, R. A. (2019). *10 rainforest facts for 2020*. Retrieved from Mongabay: https://rainforests.mongabay.com/facts/rainforest-facts.html

Cal Recycle (2020). *Compost and mulch use in agriculture: organic materials management*. Retrieved from Cal Recycle: www.calrecycle.ca.gov/organics/ farming

Calafat, A. (2008). Exposure of the U.S. population to bisphenol A and 4-tertiary-octylphenol: 2003–2004. *Environmental Health Perspectives*, **116**, 39–44.

Campbell, T. & Campbell, T. (2016). *The China Study*. Dallas, TX: BenBella books.

Campion, A. L., Oury, F., Heumez, E., & Rolland, B. (2020). Conventional versus organic farming systems: dissecting comparisons to improve cereal organic breeding strategies. *Organic Agriculture*, **10**, 63–74.

Cardona, O., Aalst, M. V., Birkmann, J., Fordham, M., McGregor, G., Perez, R., ... Sinh, B. (2020). Determinants of risk: exposure and vulnerability. In C. V.-K. Field (ed.), *Managing the Risks of Extreme Events and Disasters to Advance Climate Change Adaptation*. Cambridge and New York: Cambridge University Press, pp. 65–108.

CARP & CEPA (2011). *Method for Estimating Greenhouse Gas Emission Reduction from Compost from Commercial Organic Waste*. Sacramento: California Environment Protection Agency.

Carrero, J., Gonzalez-Ortiz, A., Avesani, C., Bakker, S., Bellizzi, V., Chauveau, P., ... Fouque, D. (2020). Plant-based diets to manage the risks and complications of chronic kidney disease. *Nature Reviews Nephrology*, 16, 525–542.

Castro, J. (2013). *Does leaving fluorescent lights on save energy?* Retrieved from LiveScience: www.livescience.com/38355-fluorescent-lights-save-energy.html

CDC (2005). *Severe acute respiratory syndrome.* Retrieved from Centers for Disease Control and Prevention: www.cdc.gov/sars/about/faq.html

CDC (2016). *Adult obesity facts.* Retrieved from Centers for Disease Control and Prevention: www.cdc.gov/obesity/data/adult.html

CDC (2017). *Zoonotic diseases.* Retrieved from Centers for Disease control and prevention: www.cdc.gov/onehealth/basics/zoonotic-diseases.html

CDC (2019a). *Antibiotic resistance and food safety.* Retrieved from Centers for Disease Control and Prevention: www.cdc.gov/foodsafety/challenges/antibiotic-resistance.html

CDC (2019b). *Reconstruction of the 1918 influenza pandemic virus.* Retrieved from Centers for Disease Control and Prevention: www.cdc.gov/flu/about/qa/1918flupandemic.htm

CDC (2020). *Adult obesity facts.* Retrieved from Centers for Disease Control and Prevention: www.cdc.gov/obesity/data/adult.html

CDP (2020). *2019–2020 Australian bushfires.* Retrieved from Center for Disaster Philanthropy: https://disasterphilanthropy.org/disaster/2019-australian-wildfires/

Ceballos, G., Ehrlich, P., & Dirzo, R. (2017). Biological annihilation via the ongoing sixth mass extinction signaled by vertebrate population losses and declines. *PNAS*, 114: E6089–E6096.

Changing Markets Foundation (2019). *Dirty Fashion Disrupted: Leaders and Laggards Revealed.* New York & London: Changing Markets Foundation.

Chase, M., Schlossberg, S., Griffin, C., Bouche, P., Djene, S., Elkan, P., ... Sutcliffe, R. (2016). Continent-wide survey reveals massive decline in African savannah elephants. *Peer J*, 4, e2354.

Chen, H. & Burns, L. (2006). Environmental analysis of textile products. *Clothing & Textiles Research Journal*, 24, 248–262.

Chen, Y., Shu, L., Qiu, Z., Lee, D., Settle, S., Hee, S., ... Allard, P. (2016). Exposure to the BPA-substitute bisphenol S causes unique alterations of germline function. *PLoS Genetics*, 12, e1006223.

Chiu, Y., Lee, W., & Chen, T. (2014). Environmentally responsible behavior in ecotourism: antecedents and implications. *Tourism Management*, 40, 321–329.

Cho, R. (2012). *What happens to all that plastic?* Retrieved from State of the Planet – Earth Institute Columbia University: https://blogs.ei.columbia.edu/2012/01/31/what-happens-to-all-that-plastic/

Christensen, L., Galt, R., & Kendall, A. (2018). Life-cycle greenhouse gas assessment of community supported agriculture in California's Central Valley. *Renewable Agriculture and Food Systems*, 33, 393–405.

CIA (2016). *China*. Retrieved from The World Factbook: www.cia.gov/library/publications/the-world-factbook/geos/ch.html

CITES (2013). *CITES Strategic Vision: 2008–2020*. Bangkok: Conference of the Parties to the Convention on International Trade in Endangered Species of Wild Fauna and Flora.

CITES (2020). *The CITES species*. Retrieved from CITES: www.cites.org/eng/disc/species.php

Clarke, C., Schweitzer, Z., & Roto, A. (2015). *Reducing Food Waste: Recommendations to the 2015 Dietary Guidelines Advisory Committee*. Boston, MA: Tufts University Friedman School of Nutrition Science and Policy.

Climate Council (2019). *Deforestation and climate change*. Retrieved from Climate Council: www.climatecouncil.org.au/deforestation/

Clover, C. (2006). *The End of the Line: How Overfishing Is Changing the World and What We Eat*. New York: The New Press.

Clover, C. (2008). *The End of the Line: How Overfishing Is Changing the World and What We Eat*. Berkeley: University of California Press.

Colarossi, N. (2020). *Photos show how cities have closed streets to cars so people have enough space to get outside during the pandemic*. Retrieved from *Insider*: www.insider.com/cities-closed-streets-for-pedestrians-covid-lockdowns-2020-5

Colbeck, M. (director) (2010). *Echo: An elephant to remember* [motion picture].

Cole, D., Cole, R., Gaydos, S., Gray, J., Hyland, G., Jacques, M., ... Au, W. (2009). Aquaculture: environmental, toxicological, and health issues. *International Journal of Hygiene and Environmental Health*, **212**, 369–377.

Compaore, A., Sirima, D., Hema, E. M., Doamba, B., Ajong, S. N., Vittorio, M. D., & Luiselli, L. (2020). Correlation between increased human–elephant conflict and poaching of elephants in Burkina Faso (West Africa). *European Journal of Wildlife Research*, **66**, 24.

Con, S. (2020). *Harnessing geothermal energy*. Retrieved from Plaza UFL: http://plaza.ufl.edu/sarahcon/geothermal.html

Cone, C. A. & Myhre, A. (2000). Community-supported agriculture: a sustainable alternative to industrial agriculture? *Human Organization*, **59**, 187–197.

Conservation International (2020). *Conservation International reports increase in poaching and tropical deforestation due to COVID-19 restrictions*. Retrieved from Conservation International: www.conservation.org/press-releases/2020/04/21/conservation-international-reports-increase-in-poaching-and-tropical-deforestation-due-to-covid-19-restrictions

CORIS (2020). *What are coral reefs?* Retrieved from Coral Reef Information System of the National Oceanic and Atmospheric Administration: www.coris.noaa.gov/about/what_are/

Cowperthwaite, G. (director). (2013). *Blackfish* [motion picture].

Cox, R., Holloway, L., Venn, L., Dowler, L., Hein, J., Kneafsey, M., & Tuomainen, H. (2008). Common ground? Motivations for participation in a community-supported agriculture scheme. *Local Environment*, 13, 203–219.

CPSC (2015). *Phthalates*. Retrieved from US Consumer Product Safety Commission: www.cpsc.gov/en/Business–Manufacturing/Business-Education/Business-Guidance/Phthalates-Information/

Cribb, J. (2019). *Food or War*. Cambridge: Cambridge University Press.

Curtin, S. (2009). Wildlife tourism: the intangible, psychological benefits of human–wildlife encounters. *Current Issues in Tourism*, 12, 451–474.

Curtin, S. & Wilkes, K. (2007). Swimming with captive dolphins: current debates and post-experience dissonance. *International Journal of Tourism Research*, 9, 131–146.

Daley, J. (2018). *Americans have a surprisingly large appetite for giraffe parts*. Retrieved from *Smithsonian Magazine*: www.smithsonianmag.com/smart-news/americans-have-surprisingly-large-appetite-giraffe-parts-180970126/

Dalton, J. (2019). *China builds dozens more "cruel, deadly" dolphin and whale theme parks*. Retrieved from The Independent: www.independent.co.uk/news/world/asia/china-dolphins-whales-marine-mammals-aquariums-captive-number-theme-park-report-a8816016.html

Daly, N. (2019). *Suffering unseen: the dark truth behind wildlife tourism*. Retrieved from National Geographic: www.nationalgeographic.com/magazine/2019/06/global-wildlife-tourism-social-media-causes-animal-suffering/

Danovich, T. (2020). *Is this the start of a CSA boom?* Retrieved from Eater: www.eater.com/2020/4/2/21200565/csa-trend-coronavirus-covid-19-stay-at-home-delivery-groceries

Dasgupta, S. (2018). *Vietnam's bear bile farms are collapsing – but it may not be good news*. Retrieved from Mongabay: https://news.mongabay.com/2018/07/vietnams-bear-bile-farms-are-collapsing-but-it-may-not-be-good-news/

Davidson, J. (2020). *Koalas face extinction threat after wildfires: new report*. Retrieved from EcoWatch: www.ecowatch.com/koalas-extinction-wildfires-australia-2645408622.html?rebelltitem=2#rebelltitem2

Davis, C. (2014). *5 things to buy in bulk to save money + reduce waste*. Retrieved from NC State University: https://sustainability.ncsu.edu/blog/changeyourstate/5-things-buy-in-bulk-save-money/

Davis, M., Faurby, S., & Svenning, J. (2018). Mammal diversity will take millions of years to recovery from the current biodiversity crisis. *PNAS*, 115, 11262–11267.

De Vos, J. D., Joppa, L., Gittleman, J., Stephens, P., & Pimm, S. (2015). Estimating the normal background rate of species extinction. *Conservation Biology*, 29, 452–462.

Deforestation Education (2020). *About palm oil.* Retrieved from Deforestation Education: www.deforestationeducation.com/products-that-contain-palm-oil.php

Denman, K., Brasseur, G., Chidthaisong, A., Ciais, P., Cox, P., Dickinson, R., . . . S. Ramachandran P. D. (2007). Couplings between changes in the climate system and biogeochemistry. In S. D. Solomon (ed.), *Climate Change 2007: The Physical Science Basis. Contribution of Working Group I to the Fourth Assessment Report of the Intergovernmental Panel on Climate Change.* Cambridge and New York: Cambridge University Press, pp. 501–568.

Department of Primary Industries and Regional Development (2018). *Composting to avoid methane production.* Retrieved from Government of Western Australia: www.agric.wa.gov.au/climate-change/composting-avoid-methane-production

Deressa, T., Ringler, C., & Hassan, R. (2010). *Factors Affecting the Choices of Coping Strategies of Climate Extremes: The Case of Farmers in the Nile Basin of Ethiopia.* Addis Ababa: EERH.

DESA, U. (2011). *World Urbanization Prospects: The 2011 Revision.* New York: United Nations Department of Economic and Social Affairs.

Desta, Y. (2014). *10 easy hacks to make your apartment more energy efficient.* Retrieved from Mashable: https://mashable.com/2014/05/14/energy-efficient-apartment/

DGAC (2015). *Part D. Chapter 4: Food Sustainability and Safety.* Washington, DC: Dietary Guidelines Advisory Committee.

Diamond, J. (2019). *Upheaval: Turning Points for Nations in Crisis.* New York: Hachette Book Group.

Diaz, R. J., & Rosenberg, R. (2008). Spreading dead zones and the consequences for marine ecosystems. *Science,* 321, 926–929.

Dolphin Project (2020a). *About us.* Retrieved from Dolphin Project: www.dolphinproject.com/about-us/history/

Dolphin Project (2020b). *Swimming with captive dolphins.* Retrieved from Dolphin Project: www.dolphinproject.com/campaigns/captivity-industry/swimming-with-dolphins/

Droogers, P., Seckler, D., & Makin, I. (2001). *Estimating the Potential of Rain-Fed Agriculture.* Washington, DC: CGIAR.

DSWT (2016). *Launching the DSWT Tsavo Dog Unit.* Retrieved from Sheldrick Wildlife Trust: www.sheldrickwildlifetrust.org/news/updates/launching-the-dswt-tsavo-dog-unit

DSWT (2020). *Water for wildlife.* Retrieved from Sheldrick Wildlife Trust: www.sheldrickwildlifetrust.org/projects/water-for-wildlife

Dunn-Emke, S., Weidner, G., & Ornish, D. (2001). Benefits of a low-fat plant-based diet. *Obesity Research,* 9, 731.

Dwyer, J., Schroeder, H., & Gobster, P. (1991). The significance of urban trees and forests: toward a deeper understanding of values. *Journal of Arboriculture*, 17, 276–284.

Earl, P. (2020). *A guide to renewable energy*. Retrieved from Compare the Market: www.comparethemarket.com/energy/content/renewable-energy-guide/

Earle, S. (2010). *The World Is Blue: How Our Fate and the Oceans Are One*. Washington, DC: National Geographic Partners LLC.

Earth Day (2018). *Fact sheet: single use plastics*. Retrieved from Earth Day: www.earthday.org/fact-sheet-single-use-plastics/

Earth Institute (2011). *Losing our coral reefs*. Retrieved from State of The Planet – Earth Institute, Columbia University: http://blogs.ei.columbia.edu/2011/06/13/losing-our-coral-reefs/

Earth Institute (2012). *What happens to all that plastic?* Retrieved from State of the Planet: http://blogs.ei.columbia.edu/2012/01/31/what-happens-to-all-that-plastic/

Earth Overshoot Day (2020a). *Earth Overshoot Day*. Retrieved from Earth Overshoot Day: www.overshootday.org/

Earth Overshoot Day (2020b). *Food demand makes up 26% of the global ecological footprint*. Retrieved from Earth Overshoot Day: www.overshootday.org/solutions/food/

Earth Policy Institute (2014). *Plastic Bags Fact Sheet*. Washington, DC: Earth Policy Institute.

EarthDay.org (2018). *Fact sheet: how much disposable plastic we use*. Retrieved from EarthDay.org: www.earthday.org/fact-sheet-how-much-disposable-plastic-we-use/

EAS (2009). Fish in–fish out ratios explained. *Aquaculture Europe*, 34. Retrieved from *Aquaculture Europe Magazine*: www.aquaeas.eu/publications-new/eas-magazine

Easterling, W., Aggarwal, P., Batima, P., Brander, K., Erda, L., Howden, S., ... Tubiello, F. N. (2007). Food, fibre and forests. In M. Parry, O. Canziani, J. Palutikof, P. V. Linden, & C. Hanson (eds.), *Climate Change 2007: Impacts, Adaptation, and Vulnerability, Contribution of Working Group II to the Fourth Assessment Report of the Intergovernmental Panel on Climate Change*. Cambridge and New York: Cambridge University Press, pp. 273–314.

EDF (2011). *Saving Lives and Reducing Health Care Costs: How Clean Air Act Rules Benefit the Nation*. New York: EDF.

Edmond, C. (2020). *How face masks, gloves, and other coronavirus waste is polluting the ocean*. Retrieved from World Economic Forum: www.weforum.org/agenda/2020/06/ppe-masks-gloves-coronavirus-ocean-pollution/

EEA (2016). *What Are the Sources of Food Waste in Europe?* Copenhagen: European Environment Agency.

EEA (2020). *Air pollution goes down as Europe takes hard measures to combat coronavirus*. Retrieved from European Environment Agency: www.eea.europa .eu/highlights/air-pollution-goes-down-as

EIA. (2013). *Today in energy*. Retrieved from EIA: www.eia.gov/todayinenergy/ detail.php?id=10271#

EIA. (2020). *Chinese government still promoting coronavirus treatment containing bear bile*. Retrieved from Environmental Investigation Agency: https://eia-international.org/news/chinese-government-still-promoting-coronavirus-treatment-containing-bear-bile/

Ellahi, M. & Rashid, M. U. (2018). The toxic effects BPA on fetuses, infants, and children. In P. Erkekoglu & B. Kocer-Gumusel (eds.), *Bisphenol A Exposure and Health Risks*. Rijeka: InTech, pp. 143–154.

Endangered Species Journalist (2020). *The status of tigers in the wild*. Retrieved from Tigers in Crisis: www.tigersincrisis.com/the-status-of-tigers/

Environment California (2016). *Keep plastic out of the Pacific*. Retrieved from Environment California: www.environmentcalifornia.org/programs/cae/keep-plastic-out-pacific

EPA (2015). *Plastics*. Washington, DC: US Environmental Protection Agency.

EPA (2016a). *Confronting plastic pollution one bag at a time*. Retrieved from US Environmental Protection Agency: https://blog.epa.gov/2016/11/01/confronting-plastic-pollution-one-bag-at-a-time/

EPA (2016b). *Nitrous oxide emissions*. Retrieved from US Environmental Protection Agency: www.epa.gov/ghgemissions/overview-greenhouse-gases

EPA (2016c). *Overview of greenhouse gases*. Retrieved from US Environmental Protection Agency: www.epa.gov/ghgemissions/overview-greenhouse-gases

EPA (2016d). *Persistent Organic Pollutants: A Global Issue, a Global Response*. Washington, DC: US Environmental Protection Agency.

EPA (2016e). *Reducing wasted food at home*. Retrieved from US Environmental Protection Agency: www.epa.gov/recycle/reducing-wasted-food-home

EPA (2017a). *Toxicological threats of plastic*. Retrieved from US Environmental Protection Agency: www.epa.gov/trash-free-waters/toxicological-threats-plastic

EPA (2017b). *Understanding global warming potentials*. Retrieved from US Environmental Protection Agency: www.epa.gov/ghgemissions/understanding-global-warming-potentials

EPA (2018). *Travel: learn more about travel assumptions in the Choose a Path tool*. Retrieved from US Environmental Protection Agency: www.epa.gov/greenvehicles/travel-learn-more-about-travel-assumptions-choose-path-tool

EPA (2019). *Fertilizer manufacturing effluent guidelines*. Retrieved from US Environmental Protection Agency: www.epa.gov/eg/fertilizer-manufacturing-effluent-guidelines

EPA (2020a). *Fast facts on transportation greenhouse gas emissions*. Retrieved from US Environmental Protection Agency: www.epa.gov/greenvehicles/fast-facts-transportation-greenhouse-gas-emissions

EPA (2020b). *Transportation sector emissions*. Retrieved from US Environmental Protection Agency: www.epa.gov/ghgemissions/sources-greenhouse-gas-emissions#transportation

Eriksen, M., Lebreton, L. C., Carson, H. S., Thiel, M., Moore, C. J., Borerro, J. C., ... Reisser, J. (2014). Plastic pollution in the world's oceans: more than 5 trillion plastic pieces weighing over 250,000 tons afloat at sea. *PLoS One, 9,* e0111913.

Esselstyn, C. (2007). Resolving the coronary artery disease epidemic through plant-based nutrition. *Preventive Cardiology, 4,* 171–177.

Essential Chemical Industry (2013). *Degradable plastics*. Retrieved from Essential Chemical Industry: www.essentialchemicalindustry.org/polymers/degradable-plastics.html

Essington, T., Moriarty, P., Froehlich, H., Hodgson, E., Koehn, L., Oken, K., ... Stawitz, C. (2015). Fishing amplifies forage fish population collapses. *PNSA, 112,* 1–5.

Estes, A. C. (2020). *America's meat shortage is more serious than your missing hamburgers: the meat supply chain is breaking down, but that's only part of the study*. Retrieved from Vox Recode: www.vox.com/recode/2020/5/8/21248618/coronavirus-meat-shortage-food-supply-chain-grocery-stores

Ethical Consumer (2020). *Why shop ethically?* Retrieved from Shop Ethical: www.ethical.org.au/3.4.2/get-informed/why-shop-ethically/

European Commission (2011). *Plastic Waste: Ecological and Human Health Impacts*. Brussels: European Commission.

European Commission (2017). *Breaking Bag Habits*. Brussels: European Commission.

European Commission (2019). *Organic Farming in the EU – A Fast Growing Sector*. Brussels: European Commission.

European Commission (2020a). *"Best Before" and "Use By" Dates on Food Packaging: Understand Them Right to Prevent Food Waste and Save Money*. Brussels: European Commission.

European Commission (2020b). *Food-based dietary guidelines in Europe*. Retrieved from European Commission of Science and Knowledge: https://ec .europa.eu/jrc/en/health-knowledge-gateway/promotion-prevention/nutrition/food-based-dietary-guidelines

European Commission (2020c). *Heating and cooling facts and figures*. Retrieved from European Commission: https://ec.europa.eu/energy/topics/energy-efficiency/heating-and-cooling_en

Eveleth, R. (2014). *"Sell by" and "best by" dates on food are basically made up but hard to get rid of.* Retrieved from *Smithsonian Magazine*: www.smithsonianmag.com/smart-news/sell-and-best-dates-food-are-basically-made-hard-get-rid-180950304/

EWG (2013). *Reduce your use of PVC in plastics and other household products.* Retrieved from Healthy Child, Healthy World – Environmental Working Group: www.healthychild.org/easy-steps/reduce-your-use-of-pvc-in-plastics-and-other-household-products/

Fair Trade Certified (2020). *Our global model.* Retrieved from Fair Trade Certified: www.fairtradecertified.org/why-fair-trade/our-global-model

FAO (1992). *Forests, Trees, and Food.* Rome: Food and Agriculture Organization of the United Nations.

FAO (2000). Livestock commodities past and present. In *World Agriculture: Towards 2015/2030.* Rome: Food and Agriculture organization of the United Nations, pp. 57–123.

FAO (2003). *On-Farm Composting Methods.* Rome: Food and Agriculture Organization of the United Nations.

FAO (2005a). *Cattle Ranching and Deforestation.* Livestock Policy Brief. Rome: Food and Agriculture Organization of the United Nations.

FAO (2005b). *Many of the world's poorest people depend on fish: fisheries and aquaculture crucial to food security, poverty alleviation.* Retrieved from Food and Agriculture Organization of the United Nations: www.fao.org/Newsroom/en/news/2005/102911/index.html

FAO (2006). *Livestock's Long Shadow: Environmental issues and options.* Rome: Food and Agriculture Organization of the United Nations.

FAO (2010). *Greenhouse Gas Emissions from the Dairy Sector: A Life Cycle Assessment.* Rome: Food and Agriculture Organization of the United Nations.

FAO (2011). *Global Food Losses and Food Waste – Extent, Causes and Prevention.* Rome: Food and Agriculture Organization of the United Nations.

FAO (2012). *World Agriculture towards 2030/2050.* Rome: Food and Agriculture Organization of the United Nations.

FAO (2013a). *Food Wastage Footprint Impacts on Natural Resources.* Rome: Food and Agriculture Organization of the United Nations.

FAO (2013b). *Food Wastage Footprint: Impacts on Natural Resources, Summary Report.* Rome: Food and Agriculture Organization of the United Nations.

FAO (2014). *The State of World Fisheries and Aquaculture. Opportunities and Challenges.* Rome: Food and Agriculture Organization of the United Nations.

FAO (2016a). *Agriculture and Food Security – The World Food Summit.* Rome: Food and Agriculture Organization of the United Nations.

FAO (2016b). *The State of World Fisheries and Aquaculture: Contributing to Food Security and Nutrition for All.* Rome: Food and Agriculture Organization of the United Nations.

FAO (2017c). *FAOSTAT*. Retrieved from Food and Agriculture Organization of the United Nations: www.fao.org/faostat/en/#data/QA

FAO (2017d). *The state of food security and nutrition in the world.* Retrieved from Food and Agriculture Organization of the United Nations: www.fao.org/state-of-food-security-nutrition/en/

FAO (2018a). *Livestock and the environment.* Retrieved from Food and Agriculture Organization of the United Nations: www.fao.org/livestock-environment/en/

FAO (2018b). *Our oceans are haunted – how "ghost fishing" is devastating our marine environments.* Retrieved from Food and Agriculture Organization of the United Nations: www.fao.org/fao-stories/article/en/c/1099596/

FAO (2018c). *The State of World Fisheries and Aquaculture.* Rome: Food and Agriculture Organization of the United Nations.

FAO (2020a). *At home in the Amazon: protecting biodiversity and livelihoods together.* Retrieved from Food and Agriculture Organization of the United Nations: www.fao.org/in-action/at-home-in-the-amazon/en/

FAO (2020b). *COVID-19 and the Risk to Food Supply Chains: How to Respond?* Rome: Food and Agriculture Organization of the United Nations.

FAO (2020c). *Food-based dietary guidelines – Japan.* Retrieved from Food and Agriculture Organization of the United Nations: www.fao.org/nutrition/education/food-dietary-guidelines/regions/countries/Japan/en

FAO (2020d). *Novel coronavirus (COVID-19) Q&A – impact on food and agriculture.* Retrieved from Food and Agriculture Organization of the United Nations: www.fao.org/2019-ncov/q-and-a/impact-on-food-and-agriculture/en/

FAO & IWMI (2017). *Water Pollution from Agriculture: A Global Review.* Rome: Food and Agriculture Organization of the United Nations.

FAO & AQUASTAT (2020). *Annual freshwater withdrawals, agriculture (% of total freshwater withdrawal).* Retrieved from The World Bank Data: https://data.worldbank.org/indicator/er.h2o.fwag.zs

FAOSTAT (2017). *FAOSTAT.* Retrieved from Food and Agriculture Organization of the United Nations: www.fao.org/faostat/en/#data/QA

Fashion Materials (2020). *Materials in fashion: spandex.* Retrieved from Fashion Materials: https://fashionmaterials.weebly.com/spandex.html

Faurby, S. & Svenning, J. (2015). Historic and prehistoric human-driven extinctions have reshaped global mammal diversity patterns. *Diversity and Distributions,* 21, 1155–1166.

Feeding America (2020). *Hunger in America.* Retrieved from Feeding America: www.feedingamerica.org/hunger-in-america/facts

Feinkel, S. (2011). A Brief History of Plastic's Conquest of the World. Retrieved from Scientific American: www.scientificamerican.com/article/a-brief-history-of-plastic-world-conquest/

Feldman, J. (2015). *Planned agricultural communities: where utopia meets suburbia*. Retrieved from Modern Farmer: https://modernfarmer.com/2015/08/planned-agricultural-communities/

Felton, A. J. & Smith, M. (2017). Integrating plant ecological responses to climate extremes from individual to ecosystem levels. *Philosophical Transactions of the Royal Society B*, 372, 20160142.

Ferreira, F., Cividanes, L., Gouveia, R., & Lona, L. (2017). An overview on properties and applications of poly(butylene adipate-co-terephthalate)–PBAT based composites. *Polymer Engineering & Science*, 59, E7–E15.

Ferrin, J., Norman, C., & Sempik, J. (2001). People, land and sustainability: community gardens and the social dimension of sustainable development. *Social Policy and Administration*, 35, 559–568.

Finco, A. D., Mamani, L. G., Carvalho, J. D., Pereira, G. D., Thomaz-Soccol, V., & Soccol, C. (2017). Technological trends and market perspectives for production of microbial oils rich in omega-3. *Critical Reviews in Biotechnology*, 37, 656–671.

Fiske, S., Crate, S., Crumley, C., Galvin, K., Lazrus, H., Lucero, L., ... Wilk, R. (2014). *Changing the Atmosphere. Anthropology and Climate Change. Final Report of the AAA Global Climate Change Task Force*. Arlington, VA: American Anthropological Association.

Flack, A. (2016). "In sight, insane": animal agency, captivity and the frozen wilderness in the late-twentieth century. *Environment and History*, 22, 629–652.

Flint, K. (2013). *Biodegradable plastic: its promises and consequences*. Retrieved from *Dartmouth Undergraduate Journal of Science*: https://sites.dartmouth.edu/dujs/2013/03/02/biodegradable-plastic-its-promises-and-consequences/

FNB, IOM, NRC, & BANR (2015). *Committee on a Framework for Assessing the Health, Environmental, and Social Effects of the Food System*. Washington, DC: National Academies Press.

Foer, J. S. (2020). *The end of meat is here*. Retrieved from *New York Times*: www.nytimes.com/2020/05/21/opinion/coronavirus-meat-vegetarianism.html?searchResultPosition=1

Folke, C. & Kautsky, N. (1992). Aquaculture with its environment: prospects for sustainability. *Ocean & Coastal Management*, 17, 5–24.

Food and Water Watch (2020). *Tap water vs. bottled water*. Retrieved from Food and Water Watch: www.foodandwaterwatch.org/about/live-healthy/tap-water-vs-bottled-water

Food Foundation (2016). *Too poor to eat: 8.4 million struggling to afford to eat in the UK*. Retrieved from The Food Foundation: https://foodfoundation.org.uk/too-poor-to-eat-8-4-million-struggling-to-afford-to-eat-in-the-uk/

FoodPrint (2020). *Community supported agriculture*. Retrieved from FoodPrint: https://foodprint.org/eating-sustainably/community-supported-agriculture/

Foot Print Network (2017). *Earth Overshoot Day*. Retrieved from Overshootday.org: www.overshootday.org/

Forgey, Q. (2020). *"Shut down those things right away": Calls to close "wet markets" ramp up pressure on China*. Retrieved from Politico: www.politico .com/news/2020/04/03/anthony-fauci-foreign-wet-markets-shutdown-162975

Hughes, J., Fothergill, A., & Scholey, K. (directors) (2020). *David Attenborough: A Life on Our Planet* [motion picture].

Francis, N., & Francis, M. J. (directors) (2006). *Black Gold* [motion picture].

Frost, R. (2020). *Should wild animals still be used for entertainment at the circus?* Retrieved from Euronews: www.euronews.com/living/2020/05/22/should-wild-animals-still-be-used-for-entertainment-at-the-circus

Fukuoka, T., Yamane, M., Kinoshita, C., Narazaki, T., Marshall, G., Abernathy, K., . . . Sato, K. (2016). The feeding habit of sea turtles influences their reaction to artificial marine debris. *Nature, 6,* 28015.

FWS (2015). *Endangered Species Act overview*. Retrieved from US Fish and Wildlife Service: www.fws.gov/endangered/laws-policies/

Gallagher, A. & Hammerschlag, N. (2011). Global shark currency: the distribution, frequency, and economic value of shark ecotourism. *Current Issues in Tourism, 14,* 797–812.

Galloway, T. S., Baglin, N., Lee, B., Kocur, A., Shepherd, M., Steele, A., & Harries, L. (2017). An engaged research study to assess the effect of a "real-world" dietary intervention on urinary bisphenol A (BPA) levels in teenagers. *BMJ, 8,* e018742.

Galt, R., Bradley, K., Christensen, L., Fake, C., Munden-Dixon, K., Sipson, N., . . . Kim, J. S. (2017). What difference does income make for community supported agriculture (CSA) members in California? Comparing lower-income and higher-income households. *Agriculture and Human Values, 34,* 435–452.

Gardiner, B. (2020). *Pollution made COVID-19 worse. Now lockdowns are clearing the air*. Retrieved from National Geographic: www .nationalgeographic.com/science/2020/04/pollution-made-the-pandemic-worse-but-lockdowns-clean-the-sky/

GCF (2018). *IUCN Red List update – giraffe*. Retrieved from Giraffe Conservation: https://giraffeconservation.org/2018/11/14/giraffe-subspecies-update/

Ge, M. & Friedrich, J. (2020). *4 charts explain greenhouse gas emissions by countries and sectors*. Retrieved from World Resources Institute: www.wri .org/blog/2020/02/greenhouse-gas-emissions-by-country-sector

Geographical Magazine (2006). *Ivory trade*. Retrieved from *Geographical Magazine*: www.geographical.co.uk/Magazine/Dossiers/Ivory_Trade_-_November_2006 .html

Gerber, P., Steinfeld, H., Henderson, B., Mottet, A., Opio, C., Dijkman, J., . . . Tempio, G. (2013). *Tackling Climate Change through Livestock: A Global*

Assessment of Emissions and Mitigation Opportunities. Rome: Food and Agriculture Organization of the United Nations.

Gerlock, G. (2014). *To end food waste, change needs to begin at home*. Retrieved from NPR: www.npr.org/sections/thesalt/2014/11/17/364172105/to-end-food-waste-change-needs-to-begin-at-home

GFAS (2020). *Global Federation of Animal Sanctuaries main page*. Retrieved from GFAS: www.sanctuaryfederation.org/accreditation/

GFN (2016). *World footprint – do we fit on the planet?* Retrieved from Global Footprint Network: www.footprintnetwork.org/en/index.php/GFN/page/world_footprint/

GHG Online (2019). *Methane sources – ruminants*. Retrieved from GHG Online: www.ghgonline.org/methaneruminants.htm

Gibbens, S. (2018). *What you need to know about plant-based plastics*. Retrieved from National Geographic: www.nationalgeographic.com/environment/2018/11/are-bioplastics-made-from-plants-better-for-environment-ocean-plastic/

Gibbens, S. (2019). *Exposed to extreme heat, plastic bottles may ultimately become unsafe*. Retrieved from National Geographic: www.nationalgeo graphic.com/environment/2019/07/exposed-to-extreme-heat-plastic-bottles-may-become-unsafe-over-time/

Gleick, P. & Cooley, H. (2009). Energy implications of bottled water. *Environmental Research Letters*, 4, 014009.

Global Commodities-Silk (2020). *Environmental impact*. Retrieved from Global Commodities-Silk: https://globalcommodities-silk.weebly.com/environmental-impact.html

Goodland, R. & Anhang, J. (2009). Livestock and climate change. *World Watch*, 22, 10–19.

Gould, H. (2015). *So, which textiles can be recycled and how?* Retrieved from *The Guardian*: www.theguardian.com/sustainable-business/sustainable-fash ion-blog/2015/feb/26/waste-recycling-textiles-fashion-industry

Gourmelon, G. (2015). *Global plastic production rises, recycling lags*. Retrieved from World Watch Institute – Vital Signs: Global Trends That Shape our Future: http://vitalsigns.worldwatch.org/vs-trend/global-plastic-production-rises-recycling-lags

GPI (2016). *Glass container recycling loop*. Retrieved from Glass Packaging Institute: www.gpi.org/recycling/glass-recycling-facts

GRAIN & IATP (2018). *Emissions Impossible: How Big Meat and Dairy Are Heating Up the Planet*. Washington, DC: GRAIN and IATP.

Green Choices (2020). *Nylon and polyester*. Retrieved from Green Choices: www .greenchoices.org/green-living/clothes/environmental-impacts

GreenMatch (2019). *The effects of paper coffee cups on the environment*. Retrieved from GreenMatch: www.greenmatch.co.uk/blog/2015/06/the-effects-of-paper-coffee-cups-on-the-environment

Grey, A. & Bolland, M. (2014). Clinical trial evidence and use of fish oil supplements. *JAMA Internal Medicine*, **174**, 460–462.

Griplas, L. (2019). *Closing the loop.* Retrieved from Woolmark: www.woolmark.com/news/interiors/closing-the-loop/

Guibourg, C. & Briggs, H. (2019). *Climate change: which vegan milk is best?* Retrieved from BBC: www.bbc.com/news/science-environment-46654042

Guillingsrud, A. (2017). *Fashion Fibers: Designing for sustainability.* New York and London: Fairchild Books, Bloomsbury.

Gunders, D. (2012). *Wasted: How America Is Losing Up to 40 Percent of Its Food from Farm to Fork to Landfill.* New York: National Resources Defense Council.

Guo, W., Pan, B., Sakkiah, S., Yavas, G., Ge, W., Zou, W., ... Hong, H. (2019). Persistent organic pollutants in food: contamination sources, health effects and detection methods. *International Journal of Environmental Research and Public Health*, **16**, 4361.

Gurian-Sherman, D. (2008). *CAFOs Uncovered – The Untold Costs of Confined Animal Feeding Operations.* Cambridge: Union of Concerned Scientists.

GWEC (2016). *Wind in numbers.* Retrieved from Global Wind Energy Council: https://gwec.net/global-figures/wind-in-numbers/

Haddad, N., Brudvig, L. A., Clobert, J., Davies, K., Gonzalez, A., Holt, R., ... Townshend, J. R. (2015). Habitat fragmentation and its lasting impact on Earth's ecosystems. *Science Advances*, **1**, e1500052.

Hale, T. (2018). *In 2017, we lost 40 football fields of trees every minute.* Retrieved from IFL Science: www.iflscience.com/environment/in-2017-we-lost-40-football-fields-of-trees-every-minute/

Hamblin, J. (2014). *Buy experiences, not things.* Retrieved from The Atlantic: www.theatlantic.com/business/archive/2014/10/buy-experiences/381132/

Handmer, J., Honda, Y., Kundzewicz, Z., Arnell, N., Benito, G., Hatfield, J., ... Yan, Z. (2012). Changes in impacts of climate extremes: human systems and ecosystems. In C. Field, V. Barros, T. Stocker, D. Qin, D. Dokken, K. Ebi, ... M. Tignor (eds.), *Managing the Risks of Extreme Events and Disasters to Advance Climate Change Adaptation.* Cambridge and New York: Cambridge University Press, pp. 231–290.

Harari, Y. N. (2015). *Sapiens: A Brief History of Humankind.* New York: Harper.

Harrington, R. (2017). *By 2050, the oceans could have more plastic than fish.* Retrieved from Business Insider: www.businessinsider.com/plastic-in-ocean-outweighs-fish-evidence-report-2017-1

Harris, E. (2009). The role of community gardens in creating healthy communities. *Australian Planner*, **46**, pp. 24–27.

Hartline, N., Bruce, N., Karba, S., Ruff, E., Sonar, S., & Holden, P. (2016). Microfiber masses recovered from conventional machine washing of new or aged garments. *Environmental Science & Technology*, **50**, 11532–11538.

Harun, S. R. & Ogneva-Himmelberger, Y. (2013). Distribution of industrial farms in the United States and socioeconomic, health, and environmental characteristics of counties. *Geography Journal*, 2013, 385893.

Harvard HSPH (2020). *The most expensive health care system in the world.* Retrieved from Harvard T.H. Chan School of Public Health: www.hsph .harvard.edu/news/hsph-in-the-news/the-most-expensive-health-care-system-in-the-world/

Harvard Medical School (2008). *Red meat and colon cancer.* Retrieved from Harvard Medical School: www.health.harvard.edu/staying-healthy/red-meat-and-colon-cancer

Harvard School of Public Health. (2016). *As overweight and obesity increase, so does the risk of dying prematurely.* Retrieved from Harvard T.H. Chan School of Public Health: www.hsph.harvard.edu/news/press-releases/overweight-obesity-mortality-risk/

Harvard School of Public Health (2020). *Linking air pollution to higher coronavirus death rates.* Retrieved from Harvard T.H. Chan School of Public Health: www.hsph.harvard.edu/biostatistics/2020/04/linking-air-pollution-to-higher-coronavirus-death-rates/

Hawkins, I. (2019). The diet, health, and environment trilemma. In J. Sabate (ed.), *Environmental Nutrition: Connecting Health and Nutrition with Environmentally Sustainable Diets.* Cambridge, MA: Academic Press, pp. 3–27.

Hawkins, I. & Sabate, J. (2013). Defining "sustainable" and "healthy" diets in an era of great environmental concern and increased prevalence of chronic diseases. *American Journal of Clinical Nutrition*, 97, 1151.

Heim, S., Stang, J., & Ireland, M. (2009). A garden pilot project enhances fruit and vegetable consumption among children. *Journal of the American Dietetic Association*, 109, 1220–1226.

Heller, P. (2006). *The Whale Warriors: Whaling in the Antarctic Sea.* New York: Simon & Schuster.

Helm, S., Serido, J., Ahn, S., Ligon, V., & Shim, S. (2019). Materialist values, financial and pro-environmental behaviors, and well-being. *Young Consumers*, 20, 264–284.

Henderson, L. & Green, C. (2020). Making sense of microplastics? Public understandings of plastic pollution. *Marine Pollution Bulletin*, 152, 110908.

Herforth, A., Arimond, M., Alvarez-Sanchez, C., Coates, J., Christianson, K., & Muehloff, E. (2019). A global review of food-based dietary guidelines. *Advances in Nutrition*, 10, 590–605.

Hermes, C., Keller, K., Nicholas, R., Segelbacher, G., & Schaefer, H. (2018). Project impacts of climate change on habitat availability for an endangered parakeet. *PLoS One*, 13, e0191773.

Herr, D. & Galland, G. (2009). *The Ocean and Climate Change.* Gland: IUCN.

Holland, K. (2020). *14 things you probably never knew about grocery store produce*. Retrieved from Reader's Digest: www.rd.com/food/things-you-never-knew-about-grocery-store-produce/

Hollenbeck, J. (2016). Interaction of the role of concentrated animal feeding operations (CAFOs) in emerging infectious diseases (EIDS). *Infection, Genetics and Evolution*, 38, 44–46.

House of Commons Environmental Audit Committee (2017). *Plastic Bottles: Turning Back the Plastic Tide*. London: House of Commons.

House of Commons Library (2019). *Obesity statistics*. Retrieved from House of Commons Library: https://commonslibrary.parliament.uk/research-briefings/sn03336/

Hsu, J. (2017). *The hard truth about the rhino horn "aphrodisiac" market*. Retrieved from Scientific American: www.scientificamerican.com/article/the-hard-truth-about-the-rhino-horn-aphrodisiac-market/

Hughes, S. A. (2017). Global sustainable farming and the SoCo Soil Conversation Project. *Denver Journal of International Law & Policy*, 45, 431–444.

Hunnes, D. (2013). The effects of weather, household assets, and safety-net programs on household food security in Ethiopia using rural household panel data. *Regional Environmental Change*, 15, 1095–1104.

Hymann, Y. (2019). *Material guide: how sustainable is hemp?* Retrieved from Good On You: https://goodonyou.eco/material-guide-hemp/

IAPF (2020). *International Anti-Poaching Foundation*. Retrieved from IAPF: www.iapf.org/

IATA (2020). *COVID-19 puts over half of 2020 passenger revenues at risk*. Retrieved from IATA: www.iata.org/en/pressroom/pr/2020-04-14-01/

IEA (2020). *Global Energy Review 2020*. Paris: International Energy Agency.

IFAD (2011). *Rural Poverty Report 2011: New Realities, New Challenges: New Opportunities for Tomorrow's Generation*. Rome: International Fund for Agricultural Development.

IFAD (2019). *Community gardens pave the way for climate-resilient agriculture in Gambia*. Retrieved from International Fund for Agricultural Development: www.ifad.org/en/web/latest/story/asset/41487388

IFPRI (2008). *Linkages between Land Management, Land Degradation, and Poverty in Sub-Saharan Africa*. Washington, DC: International Food Policy Research Institute.

IFT (2016). *The difference between "use-by," "sell-by" and "best-by" dates*. Retrieved from Institute of Food Technologists: www.ift.org/knowledge-center/learn-about-food-science/food-facts/the-difference-between-useby-sellby-and-bestby-dates.aspx

International Animal Rescue (2020). *Borneo Orangutan Rescue*. Retrieved from International Animal Rescue: www.internationalanimalrescue.org/orangutan-sanctuary?currency=USD

International Scientific Committee for Tuna and Tuna-like Species in the North Pacific Ocean (2016). *2016 Pacific bluefin tuna stock assessment.* Retrieved from International Scientific Committee for Tuna and Tuna-like species: www.iattc.org/Meetings/Meetings2016/SAC7/PDFfiles/INF/SAC-07-INF-C(a)-ISC-Letter-IATTC-Executive-Summary.pdf

IPBES (2019). *Nature's Dangerous Decline "Unprecedented"; Species Extinction Rates Accelerating.* Bonn: Intergovernmental Science Policy Platform on Biodiversity and Ecosystem Services.

IPCC (2007). *Climate Change 2007: Synthesis Report. Contribution of Working Groups I, II and III to the Fourth Assessment Report of the Intergovernmental Panel on Climate Change.* Geneva: Intergovernmental Panel on Climate Change.

IPCC (2014). *Climate Change 2014: Synthesis Report. Contribution of Working Groups I, II, and III to the Firth Assessment Report of the Intergovernmental Panel on Climate Change.* Geneva: Intergovernmental Panel on Climate Change.

IRENA (2019). *Geothermal.* Retrieved from International Renewable Energy Agency: www.irena.org/geothermal

ISC (2016). *International Scientific Committee for tuna and tuna-like species in the North Pacific Ocean.* Retrieved from Inter-American Tropical Tuna Commission: www.iattc.org/Meetings/Meetings2016/SAC7/PDFfiles/INF/SAC-07-INF-C(a)-ISC-Letter-IATTC-Executive-Summary.pdf

Islam, S., Parvin, F., Urmy, Z., Ahmed, S., Arifuzzaman, M., Yasmin, J., & Islam, F. (2020). A study on the human health benefits, human comfort properties, and ecological influences of natural sustainable textile fibers. *European Journal of Physiotherapy and Rehabilitation Studies*, 1, 1–19.

issuu (2015). *DSWT overview brochure.* Retrieved from issuu.com: https://issuu.com/davidsheldrickwildlifetrust/docs/dswt_overview_brochure_2015

IUCN (2017). *Invisible plastic particles from textiles and tyres a major source of ocean pollution – IUCN study.* Retrieved from IUCN: www.iucn.org/news/secretariat/201702/invisible-plastic-particles-textiles-and-tyres-major-source-ocean-pollution-%E2%80%93-iucn-study

Ivanova, I. (2019). *States declare war on Styrofoam.* Retrieved from CBS News: www.cbsnews.com/news/styrofoam-ban-states-declare-war-people-think-it-breaks-down/

Iwata, T. (2015). Biodegradable and bio-based polymers: future prospects of eco-friendly plastics. *Angewandte Chemie*, 54, 3210–3215.

Iyer, S. (director). (2016). *Gods in Shackles* [motion picture].

Jackson, A. (2020). *Seattle to permanently close 20 miles of streets to traffic so residents can exercise and bike on them.* Retrieved from CNN Travel: www.cnn.com/travel/article/seattle-streets-closed-stay-healthy-trnd/index.html

Jambeck, J. R., Geyer, R., Wilcox, C., Siegler, T. R., Perryman, M., Andrady, A., ... Law, K. L. (2015). Plastic waste inputs from land into the ocean. *Science*, 347, 768–771.

Jaramillo-Legorreta, A., Cardena-Hinojosa, G., Nieto-Garcia, E., Rojas-Bracho, L., Thomas, L., Hoef, J. V., ... Tregenza, N. (2019). Decline towards extinction of Mexico's vaquita porpoise (*Phocoena sinus*). *Royal Society Open Science*, 6, 190598.

JCU (2020). *Study indicates alarming fall in dolphin numbers*. Retrieved from James Cook University: www.jcu.edu.au/news/releases/2020/february/study-indicates-alarming-fall-in-dolphin-numbers

Jett, J. & Ventre, J. (2015). Captive killer whale (*Orcinus orca*) survival. *Marine Mammal Science*, 31, 1362–1377.

Ji, J. (2020). Origins of the MERS-CoV, and lessons for 2019-nCoV. *Lancet Planetary Health*, 4, E93.

Johns Hopkins University Center for a Livable Future (2016). *New U.S. dietary guidelines ignore broad support for food sustainability*. Retrieved from Johns Hopkins University HUB: http://hub.jhu.edu/2016/03/11/dietary-guidelines-sustainability-survey/

Johnson, C., Balmford, A., Brook, B., Buettel, J., Galetti, M., Guangchun, L., & Wilmshurst, J. (2017). Biodiversity losses and conservation responses in the Anthropocene. *Science*, 356, 270–275.

Johnson, J. (2020). *Arctic hits 100.4 F – hottest temperature on record*. Retrieved from Common Dreams: www.commondreams.org/news/2020/06/22/scares-me-says-bill-mckibben-arctic-hits-1004degf-hottest-temperature-record

Jones, D. W. (1991). How urbanization affects energy-use in developing countries. *Energy Policy*, 19, 621–630.

Jribi, S., Ismail, H., Doggui, D., & Debbabi, H. (2020). COVID-19 virus outbreak lockdown: what impacts on household food wastage? *Environment, Development, and Sustainability*, 22, 3939–3955.

Kahleova, H., Levin, S., & Barnard, N. (2018). Vegetarian dietary patterns and cardiovascular disease. *Progress in Cardiovascular Disease*, 61, 54–61.

Keating, J. (2018). *The sinking state*. Retrieved from *The Washington Post*: www.washingtonpost.com/news/posteverything/wp/2018/07/26/feature/this-is-what-happens-when-climate-change-forces-an-entire-country-to-seek-higher-ground/

Kelland, K. (2010). *Experts demand European action on plastics chemical*. Retrieved from Reuters: www.reuters.com/article/us-chemical-bpa-health/experts-demand-european-action-on-plastics-chemical-idUSTRE65L6JN20100622

Kelleher, K. (2005). *Discards in the world's marine fisheries. An update*. Rome: Food and Agriculture Organization of the United Nations.

Kemp, J. (2019). *Urbanization and rising energy consumption*. Retrieved from Reuters: www.reuters.com/article/us-global-energy-kemp-column-idUSKBN1XN239

Khan, J. & Arsalan, M. (2016). Solar power technologies for sustainable electricity generation – a review. *Renewable and Sustainable Energy Reviews*, **55**, 414–425.

Khare, A. & Deshmukh, S. (2006). Studies toward producing eco-friendly plastics. *Journal of Plastic Film and Sheeting*, **22**, 193–211.

Khatchadourian, R. (2007). *Neptune's Navy – Paul Watson's wild crusade to save the oceans*. Retrieved from *New Yorker*: www.newyorker.com/magazine/2007/11/05/neptunes-navy

Khokhar, T. (2013). *7 things you may not know about water*. Retrieved from World Bank Blogs: https://blogs.worldbank.org/opendata/7-things-you-may-not-know-about-water

Kim, G., Seok, J. H., Mark, T., & Reed, M. (2019). The price relationship between organic and non-organic vegetables in the US: evidence from Nielsen scanner data. *Applied Economics*, **51**, 1025–1039.

Kim, J.-Y., Bang, S., & Lee, S. (2019). Alpha-caseine changes gene expression profiles and promotes tumorigenesis of prostate cancer cells. *Nutrition and Cancer*, **72**, 239–251.

King, L., Lala, F., Nzumu, H., Mwambingu, E., & Douglas-Hamilton, I. (2017). Beehive fences as a multidimensional conflict-mitigation tool for farmers coexisting with elephants. *Conservation Biology*, **31**, 743–752.

Kinhal, V. (2020). *Types of biodegradable plastic*. Retrieved from Greenliving: https://greenliving.lovetoknow.com/Type_of_Biodegradable_Plastic

Klein, D. (2020). *Coronavirus shopping tips: what to buy and what to avoid*. Retrieved from Chowhound: www.chowhound.com/food-news/258982/coronavirus-shopping-tips-what-to-buy-what-to-avoid/

Knapp, E. J., Peace, N., & Bechtel, L. (2017). Poachers and poverty: assessing objective and subjective measures of poverty among illegal hunters outside Ruaha National Park, Tanzania. *Conservation and Society*, **15**, 24–32.

Kolbert, E. (2014). *The Sixth Extinction: An Unnatural History*. New York: Henry Holt and Company.

Koneswaran, G. & Nierenberg, D. (2008). Global farm animal production and global warming: impacting and mitigating climate change. *Environmental Health Perspectives*, **116**, 578–582.

Kosuth, M., Mason, S., & Wattenberg, E. (2018). Anthropogenic contamination of tap water, beer, and sea salt. *PLoS One*, **13**, e0194970.

Krahn, M., Hanson, M., Schorr, G., Emmons, C., Burrows, D., Bolton, J., … Ylitalo, G. (2009). Effects of age, sex and reproductive status on persistent organic pollutant concentrations in "Southern Resident" killer whales. *Marine Pollution Bulletin*, **58**, 1522–1529.

Kumar, A., Killingsworth, M., & Giilovich, T. (2014). Waiting for merlot: anticipatory consumption of experiential and material purchases. *Psychological Science*, 25, 1924–1931.

Kumar, A., Killingsworth, M., & Gilovich, T. (2020). Spending on doing promotes more moment-to-moment happiness than spending on having. *Journal of Experimental Psychology*, 88, 103971.

LA Times Editorial Board (2018). *Half a billion plastic straws are used and discarded every day. What an unacceptable waste.* Retrieved from *Los Angeles Times*: www.latimes.com/opinion/editorials/la-ed-straws-on-request-20180116-story.html

Lackner, M. (2016). PBAT: a versatile material for biodegradable and compostable packaging. *Journal of Bioremediation & Biodegradation*, 7(Suppl.).

Ladkani, R. (director). (2019). *Sea of Shadows* [motion picture].

Lahdenpera, M., Mar, K., Courtiol, A., & Lummaa, V. (2018). Differences in age-specific mortality between wild-caught and captive-born Asian elephants. *Nature Communications*, 9, 3023.

Latawiec, A., Strassberg, B., Silva, D., Alves-Pinto, H., Feltran-Barbieri, R., Castro, A., . . . Beduschi, F. (2017). Improving land management in Brazil: a perspective from producers. *Agriculture, Ecosystems & Environment*, 240, 276–286.

Latif, A. (2020). *Wildfires live updates: dozens missing as states look to weather for relief.* Retrieved from *New York Times*: www.nytimes.com/2020/09/12/us/wildfires-live-updates.html.

Lavell, A., Oppenheimer, M., Ciop, C., Hess, J., Lempert, R., Li, J., . . . Myeong, S. (2012). Climate change: new dimensions in disaster risk, exposure, vulnerability, and resilience. In *Managing the Risks of Extreme Events and Disasters to Advance Climate Change Adaptation*. Cambridge and New York: Cambridge University Press, pp. 25–64.

Layke, J. & Hutchinson, N. (2020). *3 reasons to invest in renewable energy now.* Retrieved from World Resources Institute: www.wri.org/blog/2020/05/coronavirus-renewable-energy-stimulus-packages

Leahy, S. (2019). *Climate change driving entire planet to dangerous "tipping point."* Retrieved from National Geographic: www.nationalgeographic.com/science/article/earth-tipping-point

Lenihan-Geels, G., Bishop, K. S., & Ferguson, L. R. (2013). Alternative sources of omega-3 fats: can we find a sustainable substitute for fish? *Nutrients*, 5, 1301–1315.

Lenton, T., Rockstron, J., Gaffney, O., Rahmstorf, S., Richardson, K., Steffen, W., & Schellnhuber, H. (2020). Climate tipping points – too risky to bet against. *Nature*, 575, 592–595.

Lewis Pugh Foundation (2020). *Action for oceans.* Retrieved from Lewis Pugh Foundation: https://lewispughfoundation.org/#home-banner

Lewis, S. (2019). *Beached sperm whale found with 220 pounds of trash in his stomach*. Retrieved from CBS News: www.cbsnews.com/news/beached-sperm-whale-found-with-220-pounds-of-trash-in-his-stomach/

Li, W., Tse, H., & Fok, L. (2016). Plastic waste in the marine environment: a review of sources, occurrence and effects. *Science of the Total Environment*, **566–567**, 333–349.

Lin, V. S. (2016). Research highlights: impacts of microplastics on plankton. *Environmental Science: Processes & Impacts*, **18**, 160–163.

Liu, C., Kuchma, O., & Krutovsky, K. (2018). Mixed-species versus mono-cultures in plantation forestry: development, benefits, ecosystem services and perspectives for the future. *Global Ecology and Conservation*, **15**, e00419.

Lloret, J., Ratz, H., Lleonart, J., & Demestre, M. (2016). Challenging the links between seafood and human health in the context of global change. *Journal of the Marine Biological Association of the United Kingdom*, **96**, 29–42.

Locker, M. (2019). *Australia faces "national tragedy" after koala population takes hit in recent brushfires*. Retrieved from *Smithsonian Magazine*: www.smithsonianmag.com/smart-news/australia-faces-national-tragedy-over-loss-koala-after-recent-brushfires-180973570/

Logan, T. (2019). *What really happens to the organic waste you put in your compost bin*. Retrieved from CBC-Radio Canada: www.cbc.ca/news/technology/organic-waste-composting-1.5291132

Lolu, A. J., Ahluwalia, A. S., Sidhu, M. C., Reshi, Z. A., & Mandrota, S. K. (2020). Carbon sequestration and storage by westlands: implications in the Climate Change Scenario. In A. Upadhayay, R. Singh, & D. Singh (eds.), *Restoration of Westland Ecosystem: A Trajectory Towards a Sustainable Environment*. Singapore: Springer, pp. 45–58.

Londono, E., Andreoni, M., & Casado, L. (2020). *Amazon deforestation soars as pandemic hobbles enforcement*. Retrieved from *New York Times*: www.nytimes.com/2020/06/06/world/americas/amazon-deforestation-brazil.html

LPI–OSU (2019). *Essential fatty acids*. Retrieved from Linus Pauling Institute Oregon State University: https://lpi.oregonstate.edu/mic/other-nutrients/essential-fatty-acids#reference205

Lu, J., Liu, W., Xia, R., Dai, Q., Bao, C., Tang, F., . . . Wang, Q. (2016). Effects of closing and reopening live poultry markets on the epidemic of human infection with avian influenza A virus. *Journal of Biomedical Research*, **30**, 112–119.

Lunder, S. & Sharp, R. C. (2014). *US seafood advice flawed on mercury, omega-3s*. Retrieved from Environmental Working Group: www.ewg.org/research/us-gives-seafood-eaters-flawed-advice-on-mercury-contamination-healthy-omega-3s

Lundin, J., Dills, R., Ylitalo, G., Hanson, M., Emmons, C., Schorr, G., . . . Wasser, S. (2015). Persistent organic pollutant determination in killer whale scat samples: optimization of a gas chromatography/mass spectrometry method and application to field samples. *Archives of Environmental Contamination and Toxicology*, 70, 9–19.

Lundin, J., Ylitalo, G., Booth, R., Anulacion, B., Hempelmann, J., Parsons, K., ... Wasser, S. (2016). Modulation in persistent organic pollutant concentration and profile by prey availability and reproductive status in Southern Resident killer whale scat samples. *Environmental Science & Technology*, 50, 6506–6516.

Lussier, M. (2018). *5 reasons to grow your own food*. Retrieved from University of New Hampshire: www.unh.edu/healthyunh/blog/nutrition/2018/05/5-reasons-grow-your-own-food

Macfayden, G., Huntington, T., & Cappell, R. (2009). *Abandoned, Lost or Otherwise Discarded Fishing Gear*. Rome: United Nations Environment Programme and Food and Agriculture Organization of the United Nations (UNEP & FAO).

Macias, C. J. (2020). *Is the food supply strong enough to weather COVID-19?* Retrieved from UCDavis – Feeding a Growing Population: www.ucdavis.edu/food/news/is-food-supply-strong-enough-to-weather-covid-19-pandemic/

Madigan, M. & Karhu, E. (2018). The role of plant-based nutrition in cancer prevention. *Journal of Unexplored Medical Data*, 3, 9.

Makowski, E. (2019). *Balancing palm oil and protected forests to conserve orangutans*. Retrieved from Environmental Health News: www.ehn.org/balancing-palm-oil-and-protected-forests-to-conserve-orangutans-2639221181.html?rebelltitem=1#rebelltitem1

Managa, M., Tinyani, P., Senyolo, G., Soundy, P., Sultanbawa, Y., & Sivakumar, D. (2018). Impact of transportation, storage, and retail shelf conditions on lettuce quality and phytonutrients losses in the supply chain. *Food Science & Nutrition*, 6, 1527–1536.

Marin Master Gardeners (2016). *Community gardens*. Retrieved from University of California Marin Master Gardeners: http://ucanr.edu/sites/MarinMG/Great_Gardening_Information/Marin_Community_Gardens/

Marino, L. (2020). Mental health issues in captive cetaceans. In F. D. McMillan (ed.), *Mental Health and Well-Being in Animals*, 2nd ed. Wallingford: Centre for Agriculture and Bioscience International.

Marino, L., Rose, N., Visser, I., Rally, H., Ferdowsian, H., & Slootsky, V. (2020). The harmful effects of captivity and chronic stress on the well-being of orcas (*Orcinus orca*). *Journal of Veterinary Behavior*, 35, 69–82.

Marlow, H., Harwatt, H., Soret, S., & Sabate, J. (2014). Comparing the water, energy, pesticide, and fertilizer usage for the production of foods consumed by different dietary types in California. *Public Health Nutrition*, 18, 2425–2432.

Maron, D. F. (2018a). *China legalizes rhino horn and tiger bone for medical purposes*. Retrieved from National Geographic: www.nationalgeographic.com/animals/2018/10/wildlife-watch-news-china-rhino-tiger-legal/

Maron, D. F. (2018b). *Under poaching pressure, elephants are evolving to lose their tusks.* Retrieved from National Geographic: www.nationalgeographic.com/animals/2018/11/wildlife-watch-news-tuskless-elephants-behavior-change/

Martin, M., Thottathil, S., & Newman, T. (2015). Antibiotics overuse in animal agriculture: a call to action for health care providers. *American Journal of Public Health*, 105, 2409–2410.

Martinez, S., Hand, M., Pra, M. D., Pollan, S., Ralston, K., Smith, T., ... Newman, C. (2010). *Local Food Systems: Concepts, Impacts, and Issues.* Washington, DC: USDA ERS.

Maslow, A. (1943). A theory of human motivation. *Psychological Review*, 50, 370–396.

Mastroianni, B. (2020a). *COVID-19 is causing food shortages. Here's how to manage.* Retrieved from *Healthline*: www.healthline.com/health-news/covid-19-and-food-shortages

Mastroianni, B. (2020b). *Why more people are eating plant-based protein during COVID-19.* Retrieved from *Healthline*: www.healthline.com/health-news/more-people-eating-plant-based-protein

Matthews, A. (2020). *The wild animals at risk in lockdown.* Retrieved from BBC: www.bbc.com/future/article/20200520-the-link-between-animals-and-covid-19?fbclid=IwAR1qvB-i48hkQoTJeZWq4XaQmnF1zaaJc3RNiKfAmOsHJuFEeOzAT-TsZvo

Maxwell, D. & Caldwell, R. (2008). *The Coping Strategies Index: File Methods Manual*, 2nd ed. Washington, DC: USAID.

McArthur, J. & Rasmussen, K. (2016). *Where does the world's food grow?* Retrieved from Brookings: www.brookings.edu/blog/future-development/2016/03/21/where-does-the-worlds-food-grow/

McAuley, J., & Spolar, C. (2020). *One way the coronavirus could transform Europe's cities: more space for bikes.* Retrieved from Washington Post: www.washingtonpost.com/climate-solutions/one-way-the-coronavirus-could-transform-europes-cities-more-space-for-bikes/2020/05/08/e57f2dbc-8e40-11ea-9322-a29e75effc93_story.html

McComb, K., Shannon, G., Durant, S., Sayialel, K., Slotow, R., Poole, J., & Moss, C. (2011). Leadership in elephants: the adaptive value of age. *Proceedings of the Royal Society B*, 278, 3270–3276.

McCully, P. (1996). Silenced rivers. In P. McCully (ed.), *The Ecology and Politics of Large Dams*. London: Zed Books, pp. 29–64.

MChemical (2020). *BioPBS*. Bangkok: Mistubishi Chemical.

McIntyre, D. A. & Frohlich, T. C. (2015). *10 most profitable companies in the world.* Retrieved from USA Today: www.usatoday.com/story/money/personalfinance/2015/10/24/24-7-wall-st-most-profitable-companies/74501312/

McLaughlin, O., Emmerson, M., & O'Gorman, E. (2013). Forest fragmentation. *Advances in Ecological Research*, **48**, 233–236.

McManus, L., Vasconcelos, V., Levin, S., Thompson, D., Kleypas, J., Castruccio, F., … Watson, J. (2019). Extreme temperature events will drive coral decline in the Coral Triangle. *Global Change Biology*, **26**, 2120–2133.

McPherson, E. & Rowntree, R. (1993). Energy conservation potential of urban tree planting. *Journal of Arboriculture*, **19**, 321–332.

McRae, L., Deinet, S., & Freeman, R. (2017). The Diversity-Weighted Living Planet Index: controlling for taxonomic bias in a global biodiversity indicator. *PLoS One*, **12**, e0169156.

McRae, L., Freeman, R., & Marconi, V. (2016). The Living Planet Index. In *Living Planet Report 2016: Risk and Resilience in a New Era*. Gland: WWF International, pp. 44–49.

Mederios, A. D., Maia, S. F., Santos, T. D., & Gomes, T. D. (2020). Soil carbon losses in conventional farming systems due to land-use change in the Brazilian semi-arid region. *Agriculture, Ecosystems & Environment*, **287**, 106690.

Meemken, E.-M., Sellare, J., Kouame, C., & Qaim, M. (2019). Effects of fairtrade on the livelihoods of poor rural workers. *Nature Sustainability*, **2**, 635–642.

Mehta, C., Palni, U., Franke-Whittle, I., & Sharma, A. (2014). Compost: it's role, mechanism and impact on reducing soil-borne plant diseases. *Waste Management*, **34**, 607–622.

Mendoza, H., Rubio, A. V., Garcia-Pena, G. E., Suzan, G., & Simonetti, J. A. (2020). Does land-use change increase the abundance of zoonotic reservoirs? Rodents say yes. *European Journal of Wildlife Research*, **66**, 6.

MERMAIDS EU Life+ (2016). *Report on Localization and Estimation of Laundry Microplastics Sources and on Micro and Nanoplastics Present in Washing Wastewater Effluents*. MERMAIDS EU Life+.

Mersha, M., Patel, B., Patel, D., Richardson, B., & Dhillon, H. (2015). Effects of BPA and BPS exposure limited to early embryogenesis persist to impair non-associative learning in adults. *Behavioral and Brain Health*, **11**, 27.

Mexicanist (2020). *Mexico seizes more than 15,000 turtles that would be illegally sent to China*. Retrieved from Mexicanist: www.mexicanist.com/l/illegal-turtle-trade/

Milazzo, M., Cattano, C., Alonzo, S. H., Foggo, A., Gristina, M., Rodolfo-Metalpa, R., … Hall-Spencer, J. M. (2016). Ocean acidification affects fish spawning but not paternity at CO_2 seeps. *Proceedings of the Royal Society of London B*, **283**, 20161021.

Mishra, S., Xu, J., Agarwal, U., Gonzales, J., Levin, S., & Barnard, N. (2013). A multicenter randomized controlled trial of a plant-based nutrition program to reduce body weight and cardiovascular risk in the corporate setting: the GEICO study. *European Journal of Clinical Nutrition*, **67**, 718–728.

Montgomery, S. L. (2017). *Tax bill-approved Arctic drilling will plunder 10 billion barrels of oil and destroy on million acres of wilderness.* Retrieved from Newsweek: www.newsweek.com/tax-bill-approved-arctic-drilling-will-plunder-10-billion-barrels-oil-and-753670

Montoya, F. G., Pena-Garcia, A., Juaidi, A., & Manzano-Agugliaro, F. (2017). Indoor lighting techniques: an overview of evolution and new trends for energy saving. *Energy and Buildings,* **140,** 50–60.

Morell, V. (2015). *Inside the fight to stop giraffes' "silent extinction."* Retrieved from National Geographic: www.nationalgeographic.com/news/2015/06/150625-giraffes-animals-science-conservation-africa-endangered/

Moss, G. (2015). *Here's how much money you could save by making coffee at home.* Retrieved from Business Insider: www.businessinsider.com/heres-how-much-money-you-could-save-by-making-coffee-at-home-2015-10

Mostafa, N., Farag, A., Abo-dief, H., & Tayeb, A. (2018). Production of biodegradable plastic from agricultural wastes. *Arabian Journal of Chemistry,* **11,** 546–553.

Mottet, A., Haan, C. D., Falcucci, A., Tempio, G., Opio, C., & Gerber, P. (2017). Livestock: on our plates or eating at our table? A new analysis of the feed/food debate. *Global Food Security,* **14,** 1–8.

Moyle, B. (2009). The black market in China for tiger products. *Global Crime,* 10, 124–143.

Muncke, J., Myers, J. P., Scheringer, M., & Porta, M. (2014). Food packaging and migration of food contact materials: will epidemiologists rise to the neotoxic challenge? *Journal of Epidemiology and Community Health,* **68,** 592–594.

Myers, R. A., Hutchings, J. A., & Barrowman, N. J. (1997). Why do fish stocks collapse? The example of cod in Atlantic Canada. *Environmental Applications,* **7,** 91–106.

Napper, I. & Thompson, R. (2016). Release of synthetic microplastic plastic fibres from domestic washing machines: effects of fabric type and washing conditions. *Marine Pollution Bulletin,* **112,** 39–45.

NASA (2007). *Temporary drought or permanent desert?* Retrieved from Earth Observatory NASA: https://earthobservatory.nasa.gov/features/Desertification/desertification2.php

Naser, M., Swapan, M., Ahsan, R., Afroz, T., & Ahmed, S. (2019). Climate change, migration and human rights in Bangladesh: perspectives on governance. *Asia Pacific Viewpoint,* **60,** 175–190.

National Conference of State Legislatures (2020). *State plastic and paper bag legislation.* Retrieved from National Conference of State Legislatures: www.ncsl.org/research/environment-and-natural-resources/plastic-bag-legislation.aspx

National Geographic (2009). *Plastic breaks down in ocean, after all – and fast.* Retrieved from National Geographic News: http://news.nationalgeographic.com/news/2009/08/090820-plastic-decomposes-oceans-seas.html

National Geographic (2020a). *Humus*. Retrieved from National Geographic Resource Library Encyclopedia: www.nationalgeographic.org/encyclopedia/humus/

National Geographic (2020b). *Save the plankton; breathe freely*. Retrieved from National Geographic Resource Library: www.nationalgeographic.org/activity/save-the-plankton-breathe-freely/

Nature (1938). Chinese medicine and the pangolin. *Nature*, 141, 72.

Neilsen, C., Rahman, A., Rehman, A., Walsh, M., & Miller, C. (2017). Food waste conversion to microbial polyhydroxyalkanoates. *Microbial Biotechnology*, 10, 1338–1352.

Neltner, T., Kulkarni, N., Alger, H., Maffini, M., Bongard, E., Fortin, N., & Olson, E. (2011). Navigating the U.S. food additive regulatory program. *Comprehensive Reviews in Food Science and Food Safety*, 10, 342–368.

Neme, L. (2010). *Tiger farming and traditional Chinese medicine*. Retrieved from Mongabay: https://news.mongabay.com/2010/06/tiger-farming-and-traditional-chinese-medicine/

Nestle, M. (2016). *The 2015 dietary guidelines, at long last*. Retrieved from Food Politics: www.foodpolitics.com/2016/01/the-2015-dietary-guidelines-at-long-last/

Nettleton, J., Lovegrove, J., Mensink, R., & Schwab, U. (2016). Dietary fatty acids: is it time to change the recommendations? *Annals of Nutrition & Metabolism*, 68, 249-257.

Newell, P. (2005). Race, class and the global politics of environmental inequality. *Global Environmental Politics*, 5, 70–95.

NHES (2020). *Circuses and road shows*. Retrieved from the National Humane Education Society: www.nhes.org/animal-info-2/animal-welfare-information/circuses-and-road-shows/

NIEHS–NIH (2016). *Endocrine Disruptors*. Retrieved from National Institute of Environmental Health Sciences–NIH: www.niehs.nih.gov/health/topics/agents/endocrine/

NIH–ODS (2019). *Omega-3 fatty acids: a fact sheet for health professionals*. Retrieved from National Institutes of Health Office of Dietary Supplements: https://ods.od.nih.gov/factsheets/Omega3FattyAcidsandHealth-HealthProfessional/

NOAA (2008). *Importance of Coral Reefs*. National Oceanic and Atmospheric Administration.

NOAA (2014). *How has the ocean made life on land possible*. Retrieved from National Oceanic and Atmospheric Administration: https://oceanexplorer.noaa.gov/facts/oceanproduction.html

NOAA (2016). *What is ghostfishing?* Retrieved from National Oceanic and Atmospheric Administration – National Ocean Service: http://oceanservice.noaa.gov/facts/ghostfishing.html

NOAA (2020a). *A guide to plastic in the ocean.* Retrieved from National Oceanic and Atmospheric Administration: https://oceanservice.noaa.gov/hazards/mari nedebris/plastics-in-the-ocean.html

NOAA (2020b). *How much oxygen comes from the ocean?* Retrieved from National Oceanic and Atmospheric Administration – National Ocean Service: https://oceanservice.noaa.gov/facts/ocean-oxygen.html

NOAA-Fisheries (2016). *Feeds for aquaculture.* Retrieved from National Oceanic and Atmospheric Administration Fisheries: www.nmfs.noaa.gov/aquaculture/ faqs/faq_feeds.html

Normile, D. (2016). El Nino's warmth devastating reefs worldwide. *Science, 352,* 15–16.

Nowak, D., Hoehn, R., & Crane, D. (2007). Oxygen production by urban trees in the United States. *Arboriculture & Urban forestry, 33,* 220–226.

NPR (2014). *In Europe, ugly sells in the produce aisle.* Retrieved from National Public Radio – KCRW: www.npr.org/sections/thesalt/2014/12/09/ 369613561/in-europe-ugly-sells-in-the-produce-aisle

NPR (2016). *Wal-Mart, America's largest grocer, is now selling ugly fruit and vegetables.* Retrieved from National Public Radio – KCRW: www.npr.org/ sections/thesalt/2016/07/20/486664266/walmart-world-s-largest-grocer-is-now-selling-ugly-fruit-and-veg

Nuwer, R. (2018). *Sudan, the last male northern white rhino, dies in Kenya.* Retrieved from *New York Times:* www.nytimes.com/2021/01/06/magazine/ the-last-two-northern-white-rhinos-on-earth.html

O'Barry, R. (2020). *Protocol for releasing captive dolphins.* Retrieved from Dolphin Project: www.dolphinproject.com/campaigns/dolphin-sanctuary-pro ject/protocol-for-releasing-captive-dolphins/

Obbard, R., Sadri, S., Wong, Y., Khitun, A., Baker, I., & Thompson, R. (2014). Global warming releases microplastic legacy frozen in Arctic sea ice. *Earth's Future, 2,* 315–320.

Oceana (2020). *Shark fin trade: why it should be banned in the United States.* Retrieved from Oceana: https://usa.oceana.org/publications/reports/shark-fin-trade-why-it-should-be-banned-united-states

O'Connell-Rodwell, C. (2010). *How male elephants bond.* Retrieved from *Smithsonian Magazine:* www.smithsonianmag.com/science-nature/how-male-elephants-bond-64316480/

OEHHA (2009). *Toxicological Profile for Bisphenol A.* Sacramento, CA: Office of Environmental Health Hazard Assessment.

Oever, M. V., Molenveld, K., Zee, M. V., & Bos, H. (2017). *Bio-based and Biodegradable Plastics – Facts and Figures.* Wageningen: Wageningen Food & Biobased Research.

Okin, G. (2017). Environmental impacts of food consumption by dogs and cats. *PLoS One, 12,* e0181301.

O'Malley, M., Lee-Brooks, K., & Medd, H. (2013). The global economic impact of manta ray watching tourism. *PLoS One*, 8, e65051.

O'Neil, D. (2014). *Early modern Homo sapiens*. Retrieved from Evolution of Modern Humans: http://anthro.palomar.edu/homo2/default.htm

Ontl, T. A. & Schulte, L. A. (2012). Soil carbon storage. *Nature Education Knowledge*, 3, 35.

Orca Network (2020). *Lolita's capture*. Retrieved from Orca Network: www.orcanetwork.org/Main/index.php?categories_file=Lolitas%20Capture

Ozen, S. & Darcan, S. (2011). Effects of environmental endocrine disruptors on pubertal development. *Journal of Clinical Research in Pediatric Endocrinology*, 3, 1–6.

Palansooriya, K., Shaheen, S., Chen, S., Tsang, D., Hashimoto, Y., Hou, D., . . . Ok, Y. (2020). Soil amendments for immobilization of potentially toxic elements in contaminated soils: a critical review. *Environment International*, 134, 105046.

Paris, F. (2019). *Threatened bluefin tuna sells for $3 million in Tokyo market*. Retrieved from National Public Radio: www.npr.org/2019/01/05/682526465/threatened-bluefin-tuna-sells-for-5-000-per-pound-in-tokyo-market

Parker, L. (2018). *Fast facts about plastic pollution*. Retrieved from National Geographic: www.nationalgeographic.com/news/2018/05/plastics-facts-infographics-ocean-pollution/

Patton, D. (2020). *Special report: before coronavirus, China bungled swine epidemic with secrecy*. Retrieved from Reuters: www.reuters.com/article/us-swinefever-china-epidemic-specialrepo/special-report-before-coronavirus-china-bungled-swine-epidemic-with-secrecy-idUSKBN20S189

PAWS (2020). *About PAWS*. Retrieved from Performing Animal Welfare Society: www.pawsweb.org/

Payam, D., Nieuwenhuijsen, M., Esnaola, M., Forns, J., Basagana, X., Alvarez-Pedrerol, M., . . . Sunyer, J. (2015). Green spaces and cognitive development in primary schoolchildren. *PNAS*, 112, 7937–7942.

Pchelin, P. & Howell, R. (2014). The hidden cost of value-seeking: people do not accurately forecast the economic benefits of experiential purchases. *Journal of Positive Psychology*, 9, 322–334.

Pearce, F. (2019). *As climate change worsens, a cascade of tipping points looms*. Retrieved from Yale Environment 360: https://e360.yale.edu/features/as-climate-changes-worsens-a-cascade-of-tipping-points-looms

Peltomaa, E., Johnson, M., & Taipale, S. (2018). Marine cryptophytes are great sources of EPA and DHA. *Marine Drugs*, 16, 3.

Perch-Neilsen, S., Battig, M., & Imboden, D. (2008). Exploring the link between climate change and migration. *Climatic Change*, 91, 375–393.

Perry, F. (2020). *How cities are clamping down on cars*. Retrieved from Future Planet BBC: www.bbc.com/future/article/20200429-are-we-witnessing-the-death-of-the-car

Peters, G., Andrew, R., Canadell, J., Friedlingstein, P., Jackson, R., Korsbakken, J., ... Peregon, A. (2019). Carbon dioxide emissions continue to grow despite emerging climate policies. *Nature Climate Change*, 10, 3–6.

Petre, A. (2018). *What is BPA and why is it bad for you?* Retrieved from *Healthline*: www.healthline.com/nutrition/what-is-bpa

Phalan, B., Green, R., Dicks, L., Dotta, G., Feniuk, C., Lamb, A., ... Balmford, A. (2016). How can higher-yield farming help to spare nature? *Science*, 351, 450–451.

Pimentel, D. & Pimentel, M. (2003). Sustainability of meat-based and plant-based diets and the environment. *American Journal of Clinical Nutrition*, 78, 660S–663S.

Pimm, S., Jenkins, C., Abell, R., Brooks, T., Gittleman, J., Joppa, L., ... Sexton, J. (2014). The biodiversity of species and their rates of extinction, distribution, and protection. *Science*, 344, 1246752.

Pinones, A. & Fedorov, A. V. (2016). Projected changes of Antarctic krill habitat by the end of the 21st century. *Geophysical Research Letters*, 43, 8580–8589.

Pistollato, F. & Battino, M. (2014). Role of plant-based diets in the prevention and regression of metabolic syndrome and neurodegenerative diseases. *Trends in Food Science & Technology*, 40, 62–81.

Plastic Soup Foundation (2018). *California legislation will require polyester clothing to have a microfiber pollution label.* Retrieved from Plastic Soup Foundation: www.plasticsoupfoundation.org/en/2018/03/california-legislation-will-require-polyester-clothing-to-have-a-microfiber-pollution-label/?gclid=EAIa IQobChMIq82L3Yj26QIVD9bACh19awE1EAAYASAAEgL66vD_BwE

Plastics Europe (2013). *Plastics – The Facts 2013: An Analysis of European Latest Plastics Production, Demand and Waste Data.* Wemmel: Plastics Europe.

Platt, J. (2014). *Elephants are worth 76 times more alive than dead.* Retrieved from Scientific American: https://blogs.scientificamerican.com/extinction-countdown/elephants-are-worth-76-times-more-alive-than-dead-report/

Platt, J. R. (2015). *How zoos acquire endangered species.* Retrieved from Scientific American: https://blogs.scientificamerican.com/extinction-count down/how-zoos-acquire-endangered-species1/

Plumer, B. (2020). *In a first, renewable energy is poised to eclipse coal in U.S.* Retrieved from *New York Times*: www.nytimes.com/2020/05/13/climate/cor onavirus-coal-electricity-renewables.html

Pollard, S. (2019). *A Puget Sound Orca in Captivity: The Fight to Bring Lolita Home.* Charleston, SC: The History Press.

Portney, K. (2005). Civic engagement and sustainable cities in the United States. *Public Administration Review*, 65, 579–591.

Pozin, I. (2016). *The secret to happiness? Spend money on experiences, not things*. Retrieved from Forbes: www.forbes.com/sites/ilyapozin/2016/03/03/the-secret-to-happiness-spend-money-on-experiences-not-things/#20605ddc39a6

Psihoyos, L. (director) (2009). *The Cove* [motion picture].

Psihoyos, L. (director) (2015). *Racing Extinction* [motion picture].

Psihoyos, L. (director) (2018). *The Game Changers* [motion picture].

Purdy, M. (2013). *China's economy, in six charts*. Retrieved from Harvard Business Review: https://hbr.org/2013/11/chinas-economy-in-six-charts

Quammen, D. (2013). *Spillover: Animal Infections and the Next Human Pandemic*. New York: W.W. Norton & Company.

Quéré, C. L., Jackson, R., Jones, M., Smith, A., Abernethy, S., Andrew, R., ... Peters, G. (2020). Temporary reduction in daily global CO_2 emissions during the COVID-19 forced confinement. *Nature Climate Change*, 10, 647–653.

Rashid, B. (2013). Post visit assessment: the influence of consumption emotion on tourist future intention. *IOSR Journal of Business and Management*, 9, 39–45.

Reardon, T., Bellemare, M., & Zilberman, D. (2020). *How COVID-19 may disrupt food supply chains in developing countries*. Retrieved from International Food Policy Research Institute: www.ifpri.org/blog/how-covid-19-may-disrupt-food-supply-chains-developing-countries

Reid, A. & Budge, S. (2014). Identification of unresolved complex mixtures (UCMs) of hydrocarbons in commercial fish oil supplements. *Advanced Science*, 95, 423–428.

Reid, C. (2020). *Paris to create 650 kilometers of post-lockdown cycleways*. Retrieved from Forbes: www.forbes.com/sites/carltonreid/2020/04/22/paris-to-create-650-kilometers-of-pop-up-corona-cycleways-for-post-lockdown-travel/#5e48badf54d4

Reinicke, C. (2019). *These 3 lawsuits are protecting the rights of companies like Beyond Meat to call their products "burgers," "hot dogs," and other words associated with meat*. Retrieved from Markets Insider: https://markets.businessinsider.com/news/stocks/what-states-have-laws-what-to-call-plant-based-meat-2019-7-1028436300

Reiss, D. (2011). *The Dolphin in the Mirror: Exploring Dolphin Minds and Saving Dolphin Lives*. Boston, MA, and New York: Houghton Mifflin Harcourt.

Resnick, B. (2019). *More than ever, our clothes are made of plastic. Just washing them can pollute the oceans*. Retrieved from Vox: www.vox.com/the-goods/2018/9/19/17800654/clothes-plastic-pollution-polyester-washing-machine

Reuters (2020). *Discarded coronavirus masks clutter Hong Kong beaches, trails*. Retrieved from Reuters: www.reuters.com/article/us-health-coronavirus-hongkong-environme/discarded-coronavirus-masks-clutter-hong-kongs-beaches-trails-idUSKBN20Z0PP

Rice, A., Sacco, L., Hyder, A., & Black, R. (2000). Malnutrition as an underlying cause of childhood deaths associated with infectious diseases in developing countries. *Bulletin of the World Health Organization*, **78**, 1207–1221.

Richter, V. (2020). *The big five mass extinction*. Retrieved from *Cosmos Magazine*: https://cosmosmagazine.com/palaeontology/big-five-extinctions

Ritchie, H. (2019a). *Global greenhouse gas emissions from food production*. Retrieved from Our World in Data: https://ourworldindata.org/uploads/2019/11/How-much-of-GHGs-come-from-food.png

Ritchie, H. (2019b). *Humans make up just 0.01% of Earth's life – what's the rest?* Retrieved from Our World in Data: https://ourworldindata.org/life-on-earth

Ritchie, H. (2019c). *India's population growth will come to an end: the number of children has already peaked*. Retrieved from Our World in Data: https://ourworldindata.org/indias-population-growth-will-come-to-an-end

Ritchie, H. & Roser, M. (2017). *Meat and dairy production*. Retrieved from Our World in Data: https://ourworldindata.org/meat-production

Ritchie, H. & Roser, M. (2018). *Plastic pollution*. Retrieved from Our World in Data: https://ourworldindata.org/plastic-pollution

Roach, J. (2008). *Seafood may be gone by 2048, study says*. Retrieved from National Geographic: www.nationalgeographic.com/animals/2006/11/seafood-biodiversity/

Rochman, C., Kross, S., Armstron, J., Bogan, M., Darling, E., Green, S., ... Verissimo, D. (2015). Scientific evidence supports a ban on microbeads. *Environmental Science & Technology*, **49**, 10759–10761.

Rochman, C., Tahir, A., Williams, S., Baxa, D., Lam, R., Miller, J., ... Teh, S. (2015). Anthropogenic debris in seafood: plastic debris and fibers from textiles in fish and bivalves sold for human consumption. *Scientific Reports*, **5**, 14340.

Roff, G., Brown, C. J., Priest, M. A., & Mumby, P. J. (2018). Decline of coastal apex shark populations over the past half century. *Communications Biology*, **1**, 223.

Rojas-Bracho, L., Brusca, R., Alvarez-Borrego, S., Brownell, R., Camacho, V., Ceballow, G., ... Findley, O. V. (2019). Unsubstantiated Claims can lead to tragic conservation outcomes. *BioScience*, **69**, 12–14.

Romer, J. & Tamminen, L. (2014). Plastic bag reduction ordinances: New York City's proposed charge on all carryout bags as a model for US cities. *Tulane Environemntal Law Journal*, **27**, 237–275.

Roser-Renouf, C. M. (2016). *Global warming's six Americas and the election*. Retrieved from Yale Program on Climate Change Communication: http://climatecommunication.yale.edu/publications/six-americas-2016-election/

Royal Parks (2020). *Why are trees so important?* Retrieved from Royal Parks: www.royalparks.org.uk/parks/the-regents-park/things-to-see-and-do/gardens-and-landscapes/tree-map/why-trees-are-important

Royte, E. (2006). *Corn plastic to the rescue.* Retrieved from *Smithsonian Magazine*: www.smithsonianmag.com/science-nature/corn-plastic-to-the-rescue-126404720/

Rudee, A. (2020). *Want to help the US economy? Rethink the trillion trees act.* Retrieved from World Resources Institute: www.wri.org/blog/2020/04/corona virus-US-economic-recovery-tree-planting

Rudnik, E. (2013). Biodegradability testing of compostable polymer materials. In W. Andrew (ed.), *Handbook of Biopolymers and Biodegradable Plastics.* Amsterdam: Elsevier, pp. 213–263.

Russo, C. (2015). *The disturbing truth about where zoo animals come from.* Retrieved from The Dodo: www.thedodo.com/disturbing-truth-zoo-animal-151330581.html

Ryan, A., Keske, M., Hoffman, J., & Nelson, E. (2009). Clinical overview of algal-docosahexaenoic acid: effects on triglyceride levels and other cardiovascular risk factors. *American Journal of Therapeutics*, **16**, 183–192.

Safina, C. (2015). *Beyond Words: What Animals Think and Feel.* New York: Henry Holt and Company.

Safiq, A. (2018). *Free Lolita, the killer whale.* Retrieved from Scientific American: https://blogs.scientificamerican.com/observations/free-lolita-the-killer-whale/

Sage Publications (2008). *Compost can turn agricultural soils into a carbon sink, thus protecting against climate change.* Retrieved from Science Daily: www .sciencedaily.com/releases/2008/02/080225072624.htm

Samet, J. & Burke, T. (2020). Deregulation and the assault on science and the environment. *Annual Review of Public Health*, **41**, 347–361.

Samuel, S. (2020). *The meat we eat is a pandemic risk, too.* Retrieved from Vox: www.vox.com/future-perfect/2020/4/22/21228158/coronavirus-pandemic-risk-factory-farming-meat

Sass, C. (2014). *5 ways to eat less packaged food.* Retrieved from Health: www .health.com/nutrition/5-ways-to-eat-less-packaged-food

Sastry (2018). *What are synthetic fibres and give some examples.* Retrieved from A Plus Topper: www.aplustopper.com/synthetic-fibres-examples/

Satija, A., Bhupathiraju, S. N., Rimm, E. B., Spiegelman, D., Chiuve, S. E., Borgi, L., & Willett, W. C. (2016). Plant-based dietary patterns and incidence of type 2 diabetes in US men and women: results from three prospective cohort studies. *PLoS Medicine*, **13**, e1002039.

Savage, C. (2020). *The Trump administration's legal move to prevent a meat shortage, explained.* Retrieved from *New York Times*: www.nytimes.com/2020/04/29/us/trump-meat-shortage-coronavirus.html

Sawe, B. E. (2019). *Countries with the most low-lying urban areas.* Retrieved from WorldAtlas: www.worldatlas.com/articles/countries-with-the-most-largest-low-lying-land-urban-area.html

Sax, L. (2010). Polyethylene terephthalate may yield endocrine disruptors. *Environmental Health Perspectives*, 118, 445–448.

ScAAN (2019). *Effectiveness of Plastic Regulation around the World*. New York: Scientist Action and Advocacy Network.

Science Daily (2015). *Consumption of natural estrogens in cow's milk does not affect blood levels or reproductive health*. Retrieved from Science Daily: www .sciencedaily.com/releases/2016/08/160803124441.htm

Scientific American (2020). *Orangutans are hanging on in the same palm oil plantations that displace them*. Retrieved from Scientific American: www .scientificamerican.com/article/orangutans-are-hanging-on-in-the-same-palm-oil-plantations-that-displace-them/

Sea Shepherd (2020). *Our story – we are Sea Shepherd Conservation Society*. Retrieved from Sea Shepherd: https://seashepherd.org/our-story/

Sehgal, P. & Singh, N. (2010). Impact of eco-friendly products on consumer behavior. *CBS E-Journal Biz n Bytes*, 6, 1–16.

Serdeczny, O., Adams, S., Baarsch, F., Coumou, D., Robinson, A., Hare, W., ... Perrete, M. (2016). Climate change impacts in sub-Saharan Africa: from physical changes to their social repercussions. *Regional Environmental Change*, 17, 1585–1600.

Serna, J. (2020). *Meet the California firefighters helping Australia battle epic bush fires*. Retrieved from *Los Angeles Times*: www.latimes.com/environment/story/2020-01-21/us-firefighters-australia-wildfires-adapt

Seufert, V., Ramankutty, N., & Foley, J. (2012). Comparing the yields of organic and conventional agriculture. *Nature*, 485, 229–232.

SFP (2018). *Reduction fisheries: SFP fisheries sustainability overview 2018*. Retrieved from Sustainable Fisheries Partnership: www.sustainablefish.org/ News/SFP-releases-2018-reduction-fisheries-report

Shaffer, L., Khadka, K., Hoek, J. V., & Naithani, K. (2019). Human–elephant conflict: a review of current management strategies and future directions. *Frontiers in Ecology and Evolution*, 6, 235.

Sheldrick Wildlife Trust (2020). *Mission & history*. Retrieved from Sheldrick Wildlife Trust: www.sheldrickwildlifetrust.org/about/mission-history

Shepherdson, D., Carlstead, K., & Wielebnowski, N. (2004). Cross-institutional assessment of stress responses in zoo animals using longitudinal monitoring of faecal corticoids and behavior. *Animal Welfare*, 13, S105–S113.

Shepon, A., Eshel, G., Noor, E., & Milo, R. (2016). Energy and protein feed-to-food conversion efficiencies in the US and potential food security gains from dietary changes. *Environmental Research Letters*, 11, 105002.

Shiffman, D. (2020). *Sharks*. Retrieved from *Smithsonian Magazine*: https://ocean .si.edu/ocean-life/sharks-rays/sharks

Shultz, D. (2014). *How much plastic is there in the ocean?* Retrieved from *Science Magazine*: www.sciencemag.org/news/2014/12/how-much-plastic-there-ocean

Siewerski, M. (director) (2020). *Takeout* [motion picture].

Simon, J. (2020). *How to compost at home.* Retrieved from National Public Radio: www.npr.org/2020/04/07/828918397/how-to-compost-at-home

Simon, D., Davies, G., & Ancrenaz, M. (2019). Changes to Sabah's orangutan population in recent times: 2002–2017. *PLoS One,* **14,** e0218819.

Sims, C. & Palikhe, H. (2019). Proposed changes would increase the cost and decrease the benefit of listing species as endangered. *Choices,* **34,** 1–10.

Siscovick, D., Barringer, T., Fretts, A., Wu, J., Lichtenstein, A., Costello, R., . . . Mozaffarian, D. (2017). Omega-3 polyunsaturated fatty acid (fish oil) supplementation and the prevention of clinical cardiovascular disease: a science advisory from the American Heart Association. *Circulation,* **135,** e1–e18.

Slavikova, S. P. (2019). *Advantages and disadvantages of monoculture farming.* Retrieved from Green Tumble: https://greentumble.com/advantages-and-disad vantages-of-monoculture-farming/#biodiversity

Smith, M. & Meade, B. (2019). *Who are the world's food insecure? Identifying the risk factors of food insecurity around the world.* Retrieved from USDA-ERS: www.ers.usda.gov/amber-waves/2019/june/who-are-the-world-s-food-insecure-identifying-the-risk-factors-of-food-insecurity-around-the-world/

Spaderna, H., Zahn, D., Pretsch, J., Connor, S., Zittermann, A., Schleithoff, S., . . . Weidner, G. (2013). Dietary habits are related to outcomes in patients with advanced heart failure awaiting heart transplantation. *Journal of Cardiac Failure,* **19,** 240–250.

SPI (2015). *Economic statistics.* Retrieved from the Plastics Industry Trade Association: www.plasticsindustry.org/AboutPlastics/content.cfm?ItemNumber= 658

SPI (2016a). *FFBD environmental issues – history of the plastic bag.* Retrieved from the Plastics Industry Trade Association: www.plasticsindustry.org/ IndustryGroups/content.cfm?ItemNumber=521

SPI (2016b). *History of plastics.* Retrieved from the Plastics Industry Trade Association: www.plasticsindustry.org/AboutPlastics/content.cfm?ItemNumber= 670

Sprague, M., Betancor, M., & Tocher, D. (2017). Microbial and genetically engineered oils as replacements for fish oil in aquaculture feeds. *Biotechnology Letters,* **39,** 1599–1609.

Sprague, M., Dick, J., & Trocher, D. (2016). Impact of sustainable feeds on omega-3 long-chain fatty acid levels in farmed Atlantic salmon, 2006–2015. *Scientific Reports,* **6,** 21892.

Springmann, M., Godfray, H. C., Rayner, M., & Scarborough, P. (2016). Analysis and Valuation of the health and climate change cobenefits of dietary change. *PNAS,* **113,** 4146–4151.

Stancil, J. M. (2019). *The power of one tree – the very air we breate*. Retrieved from USDA: www.usda.gov/media/blog/2015/03/17/power-one-tree-very-air-we-breathe

Statista (2016). *Total U.S. resin production from 2008 to 2015 (in million pounds)*. Retrieved from Statista: www.statista.com/statistics/203398/total-us-resin-production-from-2008/

Statista (2017). *Average annual on-the-go single-use plastics items consumed in the European Union (EU-28) as of 2017*. Retrieved from Statista: www.statista.com/statistics/815516/on-the-go-single-use-plastic-consumption-european-union-eu-28/

Steyn, P. (2016). *African elephant numbers plummet 30 percent, landmark survey finds*. Retrieved from National Geographic: www.nationalgeographic.com/animals/article/wildlife-african-elephants-population-decrease-great-elephant-census

Stuart, T. (2012). The global food waste scandal. Presented at the TEDSalon Conference, London, UK.

Sullivan, H. (2020). *Australia's fire season ends, and researchers look to the next one*. Retrieved from *New York Times*: www.nytimes.com/2020/04/21/science/australia-wildfires-technology-drones.html

Sultana, B., Anwar, F., & Iqbal, S. (2008). Effect of different cooking methods on the antioxidant activity of some vegetables. *International Journal of Food Science & Technology*, 43, 560–567.

Sun, A. (director) (2014). *Plastic Paradise* [motion picture].

SWPPD (2020). *Empty freezer is an energy waster*. Retrieved from Southwest Public Power District: www.swppd.com/blog/empty-freezer-is-an-energy-waster-1

Tarkan, L. (2015). *The big business behind the local food*. Retrieved from Fortune: https://fortune.com/2015/08/21/local-food-movement-business/

Teixeira, A. A. (2015). Thermal food preservation techniques (pasteurization, sterilization, canning and blanching). In S. Bhattacharya (ed.), *Conventional and Advanced Food Processing Technologies*. Chichester: Wiley, pp. 115–128.

Tenenbaum, L. (2020). *The amount of plastic waste is surging because of the coronavirus pandemic*. Retrieved from Forbes: www.forbes.com/sites/laurate nenbaum/2020/04/25/plastic-waste-during-the-time-of-covid-19/#6b244c6c7e48

Teuten, E., Saquing, J., Knappe, D., Barlaz, M., Jonsson, S., Bjorn, A., ... Hirai. (2009). Transport and release of chemicals from plastics to the environment and to wildlife. *Philosophical Transactions of the Royal Society*, 364, 2027–2045.

Thailand Elephants (2020). *The crush*. Retrieved from Thailand Elephants: www.thailandelephants.org/the-crush

The Aluminum Association (2016). *Recycling*. Retrieved from The Aluminum Association: www.aluminum.org/industries/production/recycling

The World Counts (2020). *Transportation and tourism*. Retrieved from The World Counts: www.theworldcounts.com/challenges/consumption/transport-and-tourism/negative-environmental-impacts-of-tourism

Thornton, A. (2000). *Towards Extinction: The Exploitation of Dmall Cetaceans in Japan*. Washington, DC, and London: Emmerson Press.

Thornton, A. (2019). *This is how many animals we eat*. Retrieved from World Economic Forum: www.weforum.org/agenda/2019/02/chart-of-the-day-this-is-how-many-animals-we-eat-each-year/

Tickell, J. (2017). *Kiss the Ground*. New York: Atria/Enliven Books.

TIES (2019). *What is ecotourism?* Retrieved from The International Ecotourism Society: https://ecotourism.org/what-is-ecotourism/

Tikki Hywood Foundation (2020). *Our approach*. Retrieved from the Tikki Hywood Foundation: www.tikkihywoodfoundation.org/our-approach/

Tilman, D. & Clark, M. (2015). Food, Agriculture, & the Environment: Can We Feed the World & Save the Earth? *Daedalus*, **144**, 8–23.

Tirado, M., Hunnes, D., Cohen, M., & Lartey, A. (2015). Climate change and nutrition security in Africa. *Journal of Hunger and Environmental Nutrition*, **10**, 22–46.

Ton, S., Berry, H. L., Ebi, K., Bambrick, H., Hu, W., Green, D., ... Butler, C. D. (2016). Climate change, food, water, and population health in China. *Bulletin of the World Health Organization*, **94**, 759–765.

Toumpanakis, A., Turbull, T., & Alba-Barba, I. (2018). Effectiveness of plant-based diets in promoting well-being in the management of type 2 diabetes: a systematic review. *BMJ Open Diabetes Research & Care*, **6**, e000534.

Trasande, L., Shaffer, R., & Sathyanarayana, S. (2018). Food additives and child health. *Pediatrics*, **142**, e20181410.

Troell, M., Joyce, A., Chopin, T., Neori, A., Buschmann, A., & Fang, J. (2009). Ecological engineering in aquaculture – potential for integrated multi-trophic aquaculture (IMTA) in marine offshore systems. *Aquaculture*, **297**, 1–9.

Troeng, S., Barbier, E., & Rodriguez, C. M. (2020). *The COVID-19 pandemic is not a break for nature – let's make sure there is one after the crisis*. Retrieved from World Economic Forum: www.weforum.org/agenda/2020/05/covid-19-coronavirus-pandemic-nature-environment-green-stimulus-biodiversity?fbclid=IwAR3q6C5_4uJpJq29kxV50X2EKzAffoeAYE2OkDt_s_71DJ_aAD46xku7xEs

Turner, J. (2019). *Shade-Grown Coffee Helps Ecosystems and Farmers*. New York: International Research Institute for Climate and Society.

Tuso, P., Ismail, M., Ha, B., & Bartolotto, C. (2013). Nutrition update for physicians: plant-based diets. *Permanente Journal*, **17**, 61–66.

UN (2015a). *We are the first generation that can end poverty, the last that can end climate change.* Retrieved from United Nations Press Release: www.un .org/press/en/2015/sgsm16800.doc.htm

UN (2015b). *World population projected to reach 9.7 billion by 2050.* Retrieved from United Nations Department of Economic and Social Affairs: www.un .org/en/development/desa/news/population/2015-report.html

UN (2019). *Desertification.* Retrieved from United Nations: www.un.org/en/ events/desertificationday/desertification.shtml

UN (2020). *Population.* Retrieved from United Nations: www.un.org/en/sections/ issues-depth/population/

UN Environment (2020). *Worldwide food waste.* Retrieved from UN Environment: www.unenvironment.org/thinkeatsave/get-informed/world wide-food-waste

UN FAO (2006). *Livestock a major threat to environment: remedies urgently needed.* Retrieved from Food and Agriculture Organization of the United Nations Newsroom: www.fao.org/newsroom/en/news/2006/1000448/index .html

UN-DESA (2015). *World population projected to reach 9.7 billion by 2050.* Retrieved from United Nations Department of Economic and Social Affairs: www.un.org/en/development/desa/news/population/2015-report.html

UN-DESA (2018). *World urbanization prospects.* Retrieved from United Nations Department of Economic and Social Affairs: https://population.un.org/wup/ Download/

UN-DESA (2019). *Growing at a slower pace, world population is expected to reach 9.7 billion and could peak at nearly 11 billion around 2100.* Retrieved from United Nations Department of Economic and Social Affairs: www.un .org/development/desa/en/news/population/world-population-prospects-2019 .html

UNEP (2013). *Microplastics.* Nairobi: United Nations Environment Programme.

UNEP (2014). *Plastic waste causes financial damage of us$13 billion to marine ecosystems each year as concern grows over microplastics.* Retrieved from United Nations Enviornment Programme News Centre: www.unep.org/news centre/default.aspx?DocumentID=2791&ArticleID=10903

UNEP (2015). *Biodegradable Plastics & Marine Litter: Misconceptions, Concerns, and Impacts on Marine Environments.* Nairobi: United Nations Environmen Programme (UNEP).

UNEP (2016a). *Marine Plastic Debris and Microplastics, Global Lessons and Research to Inspire Action and Guide Policy Change.* Nairobi: United Nations Environment Programme.

UNEP (2016b). *Plastic and microplastics in our oceans – a serious environmental threat.* Retrieved from United Nations Environment Programme: www.unep .org/stories/Ecosystems/Plastic-and-Microplastics-in-our-Oceans.asp

UNEP (2016c). *Tourism's Impact on Reefs: An Ecosystem under Threat.* Nairobi: United Nations Environment Programme.

UNEP (2018). *Supporting a Shift to More Sustainble Fishing.* Nairobi: United Nations Environmental Program.

UNEP (2019). Emissions Gap Report 2019 – global progress report on climate action. Retrieved from United Nations Environment Programme: www .unenvironment.org/interactive/emissions-gap-report/2019/

UNFCCC (2015). *Paris Agreement to the United Nations Framework Convention on Climate Change.* Rio de Janeiro and New York: United Nations.

United States Census Bureau (2016). *World population.* Retrieved from United States Census Bureau International Data Base: www.census.gov/population/ international/data/worldpop/table_population.php

University of Wyoming (2020). *CSA and composting programs.* Retrieved from University of Wyoming ACRES student farm: www.uwyo.edu/uwacres/csa-composting/index.html

UNL (2020). *Pathogens and organic matter.* Retrieved from University of Nebraska – Lincoln Institute of Agriculture and Natural Resources: https://water.unl.edu/article/animal-manure-management/pathogens-and-organic-matter

UNRIC (2013). *100 Billion plastic bags used annually in the US.* Retrieved from United Nations Regional Information Centre: www.unric.org/en/latest-un-buzz/28776-100-billion-plastic-bags-used-annually-in-the-us

UNSCN (2018). *Non-communicable Diseases, Diets, and Nutrition.* Geneva: United Nations.

US Department of Energy (2012). *Program your thermostat for fall and winter savings.* Retrieved from Energy Saver – US Department of Energy: www.energy.gov/energysaver/articles/program-your-thermostat-fall-and-winter-savings

US DHHS & USDA (2015). *2015–2020 Dietary Guidelines for Americans*, 8th ed. Washington, DC: US Department of Health and Human Services and the US Department of Agriculture. Retrieved from health.gov: http://health.gov/ dietaryguidelines/2015/guidelines/

US Fish and Wildlife Service (2013). *Marine Debris: Cigarette Lighters and the Plastic Problem on Midway Atoll.* Washington, DC: US Fish and Wildlife Service.

USA Facts (2019). *Transportation is now the largest source of greenhouse gas emissions.* Retrieved from USA Facts: https://usafacts.org/articles/transporta tion-now-largest-source-greenhouse-gas-emissions/

USCC (2008). *Using Compost Can Reduce Water Pollution.* Ronkonkoma, NY: US Composting Council.

USDA (2003). *Profiling Food Consumption in America.* Washington, DC: United States Department of Agriculture.

USDA (2015). *Self-stable food safety.* Retrieved from US Department of Agriculture Food Safety and Inspection Service: www.fsis.usda.gov/wps/portal/fsis/topics/food-safety-education/get-answers/food-safety-fact-sheets/safe-food-handling/shelf-stable-food-safety/ct_index/!ut/p/a1/jZFfT8MgFMU_TR8p1M6l861pYrbqWpdFx_qy0BYoCYUGoKqfXvbnZWbTcZ8453cC915YQQwrRT4EJo50ReT-Xk

USDA (2020). *A Guide for Conventional Farmers Transitioning to Organic Certification.* Washington DC: United States Department of Agriculture.

USDA & DHHS (2015a). *Dietary Guidelines for Americans, 2015,* 8th ed. Washington, DC: US Government Printing Office.

USDA & DHHS (2015b). *Scientific Report of the 2015 Dietary Guidelines Advisory Committee: Advisory Report to the Secretary of Health and Human Services and the Secretary of Agriculture.* Washington, DC: USDA and DHHS.

USDA NRCS (2009). *Community Garden Guide Vegetable Garden Planning and Development.* Washington, DC: United States Department of Agriculture Natural Resources Conversation Service.

USDA-ERS (2016). *Food availability (per capita) data system.* Retrieved from United States Department of Agriculture – Economic Research Service: www.ers.usda.gov/data-products/food-availability-(per-capita)-data-system/.aspx#.VC1x2CtdVPQ

US-EIA (2020). *Biofuels explained.* Retrieved from US Energy Information Administration: www.eia.gov/energyexplained/biofuels/biodiesel.php

USFWS (2020). *Salmon of the west – why are salmon in trouble?* Retrieved from US Fish & Wildlife Service: www.fws.gov/salmonofthewest/dams.htm

Valdivia, S. M. & Margolies, D. (2020). *Workers sue Smithfield foods, allege conditions put them at risk for COVID-19.* Retrieved from NPR: www.npr.org/2020/04/24/844644200/workers-sue-smithfield-foods-allege-conditions-put-them-at-risk-for-covid-19

Valencia, A. (2020). *Ecauador monitoring fleet of fishing vessels near Galapagos.* Retrieved from Reuters: www.reuters.com/article/us-ecuador-fishing-idUSKCN24O369

Vally, H. (2020). *6 countries, 6 curves: how nations that moved fast against COVID-19 avoided disaster.* Retrieved from The Conversation: https://theconversation.com/6-countries-6-curves-how-nations-that-moved-fast-against-covid-19-avoided-disaster-137333

Vandy, K. (2020). *Coronavirus: how pandemic sparked European cycling revolution.* Retrieved from BBC: www.bbc.com/news/world-europe-54353914

Vann, K. (2020). *Will the grocery store bulk aisle survive COVID-19?* Retrieved from The Counter: https://thecounter.org/covid-19-coronavirus-bulk-aisle-health-grocery-beans-rice/

Vasquez, A., Sherwood, N., Larson, N., & Story, M. (2017). Community-supported agriculture as a dietary and health improvement strategy. *Journal of the Academy of Nutrition and Dietetics*, **117**, 83–94.

Veldstra, M., Alexander, C., & Marshall, M. (2014). To certify or not to certify? Separating the organic production and certification decisions. *Food Policy*, **49**, 429–436.

Vidyasagar, A. (2018). *What is photosynthesis?* Retrieved from *Live Science*: www.livescience.com/51720-photosynthesis.html

Voort, M. P., Spruijt, J., Potters, J., Wolf, P. L., & Elissen, H. (2017). *Socio-economic Assessment of Algae-based PUFA Production*. Gottingen: PUFAChain project.

Vuuren, P. J. (2017). FACTSHEET: *Africa's leading causes of death in 2016*. Retrieved from Africa Check: https://africacheck.org/factsheets/factsheet-afri cas-leading-causes-death/

Wagner, C. (2014). *Bioplastics*. Zurich: Food Packaging Forum Foundation.

Waiology (2012). *How much water does it take to produce one litre of milk?* Retrieved from Sciblogs: https://sciblogs.co.nz/waiology/2012/05/24/how-much-water-does-it-take-to-produce-one-litre-of-milk/

Waite, R. & Rudee, A. (2020). *6 ways the US can curb climate change and grow more food*. Retrieved from World Resources Institute: www.nature.com/scita ble/knowledge/library/soil-carbon-storage-84223790/

Walch, C. (2020). *COVID-19, armed conflict, and the wildlife trade*. Retrieved from The Global Observatory: https://theglobalobservatory.org/2020/05/ covid-19-armed-conflict-wildlife-trade/

Wang, S. & Ge, M. (2019). *Everything you need to know about the growing source of global emissions: transport*. Retrieved from World Resources Institute: www.wri.org/blog/2019/10/everything-you-need-know-about-fastest-growing-source-global-emissions-transport

Wang, T., Jiang, Z., Zhao, B., Gu, Y., Liou, K.-N., Kalandiyur, N., . . . Zhu, Y. (2020). Health co-benefits of achieving sustainable net-zero greenhouse gas emissions in California. *Nature Sustainability*, **3**, 597–605.

Washam, C. (2010). *Plastics Go Green*. Washington, DC: ChemMatters.

Watson, P. (2015). *If the ocean dies, we die*. Retrieved from EcoWatch: www .ecowatch.com/paul-watson-if-the-ocean-dies-we-die-1882105818.html

WDC (2020). *Why are we still putting whales and dolphins in small tanks for entertainment?* Retrieved from Whale and Dolphin Conservation: https://us .whales.org/our-4-goals/end-captivity/

Webb, H., Arnott, J., Crawford, R., & Ivanova, E. (2013). Plastic degradation and its environmental implications with special reference to poly(ethylene terephthalate). *Polymers*, **5**, 1–18.

WebMD (2018). *Fish oil – side effects*. Retrieved from WebMD: www.webmd .com/vitamins/ai/ingredientmono-993/fish-oil

West, L. (2019). *Pros and cons of PLA: corn-based plastic.* Retrieved from ThoughtCo: www.thoughtco.com/pros-cons-corn-based-plastic-pla-1203953

Westerhoff, P., Prapaipong, P., Shock, E., & Hillaireau, A. (2008). Antimony leaching from polyethylene terephthalate (PET) plastic used for bottled drinking water. *Water Research*, 42, 551–556.

WFP (2016). *Hunger statistics.* Retrieved from World Food Programme: www.wfp.org/hunger/stats

WFP (2019). *2019 – hunger map.* Retrieved from World Food Programme: www.wfp.org/publications/2019-hunger-map

WFP (2020a). *COVID-19 will double number of people facing food crises unless swift action is taken.* Retrieved from World Food Programme: www.wfp.org/news/covid-19-will-double-number-people-facing-food-crises-unless-swift-action-taken

WFP (2020b). *New report shows hunger is due to soar as coronavirus obliterates lives and livelihoods.* Retrieved from World Food Programme: www.wfp.org/news/new-report-shows-hunger-due-soar-coronavirus-obliterates-lives-and-livelihoods

White, M., Alcock, I., Grellier, J., Wheeler, B., Hartig, T., Warber, S., ... Fleming, L. (2019). Spending at least 120 minutes a week in natures is associated with good health and wellbeing. *Nature Scientific Reports*, 9, 7730.

White, R. & Hall, M. (2017). Nutritional and greenhouse gas impacts of removing animals from US agriculture. *PNAS*, 114, E10301–E10308.

Whiting, S., Berhanu, G., Haileslassie, H. A., & Henry, C. (2019). Pulses and mineral bioavilability in low income countries. In W. J. Dahl (ed.), *Health Benefits of Pulses*. Cham: Springer, pp. 43–53.

WHO (2014). *Dioxins and their effects on human health.* Retrieved from World Health Organization Media Centre: www.who.int/mediacentre/factsheets/fs225/en/

WHO (2016a). *Health statistics and information systems.* Retrieved from World Health Organization: www.who.int/healthinfo/global_burden_disease/metrics_daly/en/

WHO (2016b). *Obesity and overweight.* Retrieved from World Health Organization: www.who.int/mediacentre/factsheets/fs311/en/

WHO (2018). *Global hunger continues to rise, new UN report says.* Retrieved from World Health Organization: www.who.int/news-room/detail/11-09-2018-global-hunger-continues-to-rise—new-un-report-says

WHO (2019a). *Middle East respiratory syndrome coronavirus (MERS-CoV).* Retrieved from World Health Organization: www.who.int/emergencies/mers-cov/en/

WHO (2019b). *Middle East respiratory syndrome coronavirus (MERS-CoV).* Retrieved from World Health Organization: www.who.int/news-room/fact-sheets/detail/middle-east-respiratory-syndrome-coronavirus-(mers-cov)

WHO (2020a). *FAQs: H5N1 influenza*. Retrieved from World Health Organization: www.who.int/influenza/human_animal_interface/avian_influenza/h5n1_research/faqs/en/

WHO (2020b). *Moving around during the COVID-19 outbreak*. Retrieved from World Health Organization Regional Office for Europe: https://who.canto.global/pdfviewer/viewer/viewer.html?share=share%2Calbum%2CU6GDM&column=document&id=q3vo2qdsh114763ffe3apo4s2p&suffix=pdf

WHO (2020c). *Obesity and overweight*. Retrieved from World Health Organization: www.who.int/news-room/fact-sheets/detail/obesity-and-overweight

WHO (2020d). *WHO recommendations to reduce risk of transmission of emerging pathogens from animals to humans in live animal markets or animal product markets*. Retrieved from World Health Organization: https://apps.who.int/iris/handle/10665/332217

WHO/FAO (2002). *Diet, Nutrition, and the Prevention of Chronic Diseases*. Geneva: World Health Organization.

WHO-Euro (2020). *Data and statistics*. Retrieved from World Health Organization Regional Office for Europe: www.euro.who.int/en/health-topics/noncommunicable-diseases/obesity/data-and-statistics

Wicker, A. (2017). *Greenwashing alert: rayon viscose is made from plants, but is also toxic and destructive*. Retrieved from Ecocult: https://ecocult.com/greenwashing-alert-that-natural-fabric-made-from-plants-might-be-toxic/

Wieczorek, A., Morrison, L., Croot, P., Allcock, A., MacLoughlin, E., Savard, O., & Doyle, T. (2018). Frequency of microplastics in mesopelagic fishes from the Northwest Atlantic. *Frontiers in Marine Science, 5*, 1–9.

Wilcox, C. (2018). *Human-caused extinctions have set mammals back millions of years*. Retrieved from National Geographic: www.nationalgeographic.com/animals/2018/10/millions-of-years-mammal-evolution-lost-news/

Wong, W., Yip, Y., Choi, K., Ho, Y., & Xiao, Y. (2013). Dietary exposure to dioxins and dioxin-like PCBs of Hong Kong adults: results of the first Hong Kong Total Diet Study. *Food Additives & Contaminants, 30*, 2152–2158.

Woodyatt, A. (2020). *India's carbon emissions drop for the first time in four decades*. Retrieved from CNN: www.cnn.com/2020/05/12/india/india-carbon-emissions-fall-intl-scli/index.html

World Animal Protection (2020). *Dolphins FAQ – where do captive dolphins come from?* Retrieved from World Animal Protection: www.worldanimalprotection.ca/dolphin-faq

World Atlas (2019). *How many rhinos are there on Earth?* Retrieved from World Atlas: www.worldatlas.com/articles/how-many-rhinos-are-there-on-earth.html

World Bank (2013). *Fish to 2030, Propsects for Fisheries and Aquaculture*. Washington, DC: World Bank.

World Centric (2018). *Compostable plastics*. Retrieved from World Centric: www.worldcentric.com/blog/compostable-plastic

World Economic Forum (2020). *COVID-19 is exacerbating food shortages in Africa*. Retrieved from World Economic Forum: www.weforum.org/agenda/2020/04/africa-coronavirus-covid19-imports-exports-food-supply-chains

World Economic Forum, Ellen MacArthur Foundation, & McKinsey & Company. (2016). *The New Plastics Economy: Rethinking the Future of Plastics*. Geneva: Ellen MacArthur Foundation.

World Population Review (2020). *Bangladesh*. Retrieved from World Population Review: https://worldpopulationreview.com/countries/bangladesh-population/

Worldometer (2020). *Current world population*. Retrieved from Worldometer: www.worldometers.info/world-population/

WRAP (2020). *Food Surplus and Waste in the UK – Key Facts*. Banbury: WRAP.

Wu, J. (2016). *Shark Fin and Mobulid Ray Gill Plate Trade in Mainland China, Hong Kong, and Taiwan*. Taipei: IUCN.

WWAP (2016). *The United Nations World Water Development Report 2016: Water and Jobs*. Paris: United Nations World Water Assessment Programme.

WWF (2018). *Living Planet Report – 2018: Aiming Higher*. Gland: WWF.

WWF (2020a). *Five things Tiger King doesn't explain about captive tigers*. Retrieved from World Wildlife Fund: https://blog.wwf.ca/blog/2020/04/08/5-things-tiger-king-doesnt-explain/

WWF (2020b). *Sustainable seafood, overview*. Retrieved from World Wildlife Fund: www.worldwildlife.org/industries/sustainable-seafood

Wynes, S. & Nicholas, K. (2017). The climate mitigation gap: education and government recommendations miss the most effective individual actions. *Environmental Research Letters*, 12, 074024.

Yaffe-Bellany, D. & Corkery, M. (2020). *Dumped milk, smashed eggs, plowed vegetables: food waste of the pandemic*. Retrieved from *New York Times*: www.nytimes.com/2020/04/11/business/coronavirus-destroying-food.html

Yale Environment Review (2017). *Oases in the urban "food desert?"* Retrieved from Yale Environment Review: https://environment-review.yale.edu/oases-urban-food-desert-0

Yan, S. (2020). *China touts alternative remedies from bear bile to orange peel to treat coronavirus*. Retrieved from *The Telegraph*: www.telegraph.co.uk/news/2020/04/26/china-touts-alternative-remedies-bear-bile-orange-peel-treat/

Yang, G., Kong, L., Zhao, W., Wan, X., Zhai, Y., Chen, L., & Koplan, J. (2008). Emergence of chronic non-communicable diseases in China. *Lancet*, 673, 42–50.

Yang, Y., Peng, F., Wang, R., Guan, K., Jiang, T., Xu, G., . . . Chang, C. (2020). The deadly coronaviruses: the 2003 SARS pandemic and the 2020 novel coronavirus epidemic in China. *Journal of Autoimmunity*, 109, 102434.

Yaroslavsky, Z., Burke, G. M., Knabe, D., & Antonovich, M. (2007). *An Overview of Carryout Bags in Los Angeles County*. Los Angeles, CA: County of Los Angeles Plastic Bag Working Group.

Yifan, L. (2018). *Sea turtle smugglers cashing in*. Retrieved from Chinadialogue ocean: https://chinadialogueocean.net/4096-smugglers-cashing-in-on-sea-turtles/

Yong, E. (2018). *In a few centuries, cows could be the largest land animals left*. Retrieved from The Atlantic: www.theatlantic.com/science/archive/2018/04/in-a-few-centuries-cows-could-be-the-largest-land-animals-left/558323/

Zeller, U., Starik, N., & Gottert, T. (2017). Biodiversity, land use and ecosystem services – an organismic and comparitive approach to different geographical regions. *Global Ecology and Conservation*, 10, 114–125.

Zhong, V., Horn, L. V., & Cornelis, M. (2019). Associations of dietary cholesterol or egg consumption with incident cardiovascular disease and mortality. *JAMA*, 321, 1081–1095.

Zimmer, K. (2019). *Why the Amazon doesn't really produce 20% of the world's oxygen*. Retrieved from National Geographic: www.nationalgeographic.com/environment/2019/08/why-amazon-doesnt-produce-20-percent-worlds-oxygen/

Zong, G., Eisenberg, D. M., Hu, F. B., & Sun, Q. (2016). Consumption of meals prepared at home and risk of type 2 diabetres: an analysis of two prospective cohort studies. *PLoS Medicine*, 13, e1002052.

INDEX